Microsoft Defender for Cloud Cookbook

Protect multicloud and hybrid cloud environments, manage compliance and strengthen security posture

Sasha Kranjac

BIRMINGHAM—MUMBAI

Microsoft Defender for Cloud Cookbook

Group Product Manager: Vijin Boricha
Publishing Product Manager: Shrilekha Malpani
Senior Editor: Shazeen Iqbal
Content Development Editor: Nihar Kapadia
Technical Editor: Arjun Varma
Copy Editor: Safis Editing
Project Coordinator: Shagun Saini
Proofreader: Safis Editing
Indexer: Manju Arasan
Production Designer: Prashant Ghare
Marketing Coordinator: Hemangi Lotlikar

First published: July 2022
Production reference: 1150622

Published by Packt Publishing Ltd.
Livery Place
35 Livery Street
Birmingham
B3 2PB, UK.

978-1-80107-613-5

www.packt.com

To my family.

To Noel – for being an endless well of inspiration and love.

To Adisa – for being my loving, supporting, and encouraging partner through the highs and lows of life's journey.

To Mom and Dad – for who I am, and for where I am today.

To my grandparents – for their infinite wisdom, support, and love.

– Sasha Kranjac

Contributors

About the author

Sasha Kranjac is a security and Azure person, a cloud security architect, and an instructor. He began programming in Assembler but then met Windows NT 3.5, and was hooked on IT ever since.

He owns Kloudatech and a few other IT training and consulting companies that help companies embrace the cloud and be safe in cyberspace.

Aside from cloud and security architecture and consulting, he and his company deliver Microsoft, CompTIA, EC-Council, and their custom Azure and security courses and PowerClass workshops internationally.

Sasha is a Microsoft **Most Valuable Professional (MVP)**, **Microsoft Certified Trainer (MCT)**, MCT Regional Lead, **Certified EC-Council Instructor (CEI)**, and frequent speaker at various international conferences, events, and user groups.

Sasha has more than 100 IT certifications from Microsoft, CompTIA, and **Amazon Web Services (AWS)**, and more exams and certifications are in the queue.

About the reviewers

Rod Trent is a security CSA for Microsoft and an Azure Sentinel global SME, helping customers migrate from existing SIEMs to Azure Sentinel to achieve the promise of better security through improved efficiency without compromise.

Rod is a husband, dad, and recently a first-time grandfather. He spends his spare time (if such a thing does truly exist) simultaneously watching *Six Million Dollar Man* TV show episodes and writing KQL queries.

Table of Contents

2

Multi-Cloud Connectivity

3

Workflow Automation and Continuous Export

4

Secure Score and Recommendations

5

Security Alerts

6

Regulatory Compliance and Security Policy

7

Microsoft Defender for Cloud Workload Protection

8

Firewall Manager

Preface

Microsoft Defender for Cloud is a **Cloud Workload Protection Platform** (**CWPP**) that has **Cloud Security Posture Management** (**CSPM**) capabilities and supports Azure, on-premises, **Amazon Web Services** (**AWS**), and **Google Cloud Platform** (**GCP**) resources.

Defender for Cloud covers three crucial requirements for workload and infrastructure security: defending, securing, and continuously assessing protected workloads:

- **Defend**: Helps you detect and resolve threats to services and resources.
- **Secure**: Recommendations help you prioritize hardening tasks to improve your security posture.
- **Continuously assess**: Your secure score is frequently refreshed to give you the current security situation.

In this book, you will find valuable but easy-to-follow steps to get started using Defender for Cloud, followed by more advanced protections, including multi-cloud protection, as well as adjacent security services integrated and used in Defender for Cloud.

Who this book is for

This book is for security engineers, systems administrators, security professionals, IT professionals, system architects, developers… anyone whose responsibilities include maintaining security posture, identifying and remediating vulnerabilities, and securing cloud and hybrid infrastructure. It is also for anyone who is willing to learn about security in Azure and to build secure Azure and hybrid infrastructure, to improve their security posture in Azure, hybrid, and multi-cloud environments by using all the features within Defender for Cloud.

What this book covers

Chapter 1, *Getting Started with Microsoft Defender for Cloud*, introduces the basic but fundamental Defender for Cloud configuration and performs the initial configuration.

Chapter 2, *Multi-Cloud Connectivity*, shows you how to connect AWS and GCP environments to Defender for Cloud.

Chapter 3, *Workflow Automation and Continuous Export*, explains how to configure Defender for Cloud workflow automations, automate responses, and configure continuous data export.

Chapter 4, *Secure Score and Recommendations*, explains how to work with and interpret the secure score and manage security recommendations.

Chapter 5, *Security Alerts*, demonstrates how to manage and respond to security alerts.

Chapter 6, *Regulatory Compliance and Security Policy*, explains how to manage Defender for Cloud security policies and manage regulatory compliance standards.

Chapter 7, *Microsoft Defender for Cloud Workload Protection*, covers the protection capabilities of Defender for Cloud plans.

Chapter 8, *Firewall Manager*, demonstrates how to secure Azure assets and public endpoints by controlling network traffic to and from Azure.

Chapter 9, *Information Protection*, discusses Defender for Cloud's ability to generate alerts and recommendations based on information policy data.

Chapter 10, *Workbooks*, shows how to create and manage workbooks in Defender for Cloud.

To get the most out of this book

To successfully complete the recipes in this book, you will need an Azure subscription. Naturally, you will also need a web browser – although I have used Microsoft Edge, you can use any browser of your choice.

Additionally, for Defender for Cloud to generate alerts and recommendations, you will need to provision resources in Azure. Preferably, to create resources in AWS and GCP, you will need an account, and a payment method in these cloud providers as well.

Software/Hardware covered in the book	Requirements
Browser	Preferably Microsoft Edge, but any web browser will do
Defender for Cloud plans	Microsoft Defender for Cloud Plans
Resources	In Azure, optionally in AWS and GCP

Download the color images

We also provide a PDF file that has color images of the screenshots/diagrams used in this book. You can download it here: `https://static.packt-cdn.com/downloads/9781801076135_ColorImages.pdf`.

Conventions used

There are a number of text conventions used throughout this book.

`Code in text`: Indicates code words in text, database table names, folder names, filenames, file extensions, pathnames, dummy URLs, user input, and Twitter handles. Here is an example: "To onboard Microsoft Defender for Cloud using PowerShell, you must use the `Az.Security` PowerShell module."

A block of code is set as follows:

```
Set-AzContext -Subscription "<subscription_ID>"
Set-AzSecurityAutoProvisioningSetting '
-Name "default" -EnableAutoProvision
```

Any command-line input or output is written as follows:

```
Set-AzContext -Subscription "<subscription ID>"
```

Bold: Indicates a new term, an important word, or words that you see onscreen. For example, words in menus or dialog boxes appear in the text like this. Here is an example: "In the top menu, click **Configure**."

> **Tips or important notes**
> Appear like this.

Get in touch

Feedback from our readers is always welcome.

General feedback: If you have questions about any aspect of this book, mention the book title in the subject of your message and email us at customercare@packtpub.com.

Errata: Although we have taken every care to ensure the accuracy of our content, mistakes do happen. If you have found a mistake in this book, we would be grateful if you would report this to us. Please visit www.packtpub.com/support/errata, selecting your book, clicking on the Errata Submission Form link, and entering the details.

Piracy: If you come across any illegal copies of our works in any form on the Internet, we would be grateful if you would provide us with the location address or website name. Please contact us at copyright@packt.com with a link to the material.

If you are interested in becoming an author: If there is a topic that you have expertise in and you are interested in either writing or contributing to a book, please visit authors. packtpub.com.

Reviews

Please leave a review. Once you have read and used this book, why not leave a review on the site that you purchased it from? Potential readers can then see and use your unbiased opinion to make purchase decisions, we at Packt can understand what you think about our products, and our authors can see your feedback on their book. Thank you!

For more information about Packt, please visit packt.com.

Share Your Thoughts

Once you've read *Microsoft Defender for Cloud Cookbook*, we'd love to hear your thoughts!
Scan the QR code below to go straight to the Amazon review page for this book and share
your feedback.

https://packt.link/r/1-801-07613-8

Your review is important to us and the tech community and will help us make sure we're
delivering excellent quality content.

1

Getting Started with Microsoft Defender for Cloud

In this first chapter, you will learn how to get started with **Microsoft Defender for Cloud (MDC)** I will also introduce to you the basic but fundamental Microsoft Defender for Cloud configuration and perform initial MDC configuration steps that will set a foundation for using the program's protection and monitoring capabilities.

The recipes in this chapter will explain the essential and foundational Microsoft Defender for Cloud configuration steps that influence MDC's security capabilities, infrastructure coverage, and behavior. It is vital to know which Log Analytics Workspace will be used, the level of data that's been collected, and how monitoring agents will be deployed. Although you can change these settings anytime, it is better to set foundational and basic settings first and then proceed with configuring other settings.

After all, your choices will have an impact not only on security but on cost as well.

We will cover the following recipes in this chapter:

- Enabling Microsoft Defender for Cloud Plans on Azure Subscriptions and Log Analytics workspaces
- Enabling an Microsoft Defender for Cloud Plans on an Azure Subscription
- Enabling an Microsoft Defender for Cloud Plans on a Log Analytics workspace
- Enabling an Microsoft Defender for Cloud Plans on multiple Azure Subscriptions and Log Analytics workspaces
- Configuring data collection on a Log Analytics workspace
- Configuring provisioning extensions automatically
- Enabling a Log Analytics agent for Azure VMs manually in the Log Analytics workspace settings
- Enabling the Log Analytics agent for Azure VMs manually in the virtual machine settings
- Configuring the Log Analytics agent for Azure VMs extension deployment
- Configuring email notifications
- Assigning Microsoft Defender for Cloud permissions
- Onboarding Microsoft Defender for Cloud using PowerShell
- Enabling Microsoft Defender for Cloud integration with other Microsoft security services

Technical requirements

To complete the recipes in this chapter, the following is required:

- An Azure subscription (for some of the recipes in this chapter)
- Two or more Azure subscriptions (for some of the recipes in this chapter)
- Azure PowerShell
- A web browser, preferably Microsoft Edge

The code samples for this chapter can be found at `https://github.com/PacktPublishing/Microsoft-Defender-for-Cloud-Cookbook`.

Enabling Microsoft Defender for Cloud Plans on Azure Subscriptions and Log Analytics Workspaces

Microsoft Defender for Cloud natively protects services in Azure –no steps must be followed to enable its native, basic functionality. However, you might need to protect multiple subscriptions at a more advanced level, using **Microsoft Defender for Cloud Plans**. In the end, you will enable Microsoft Defender for Cloud Plans on multiple Azure subscriptions and **Log Analytics Workspaces** at once.

Getting ready

Before you enable Microsoft Defender for Cloud Plans on multiple subscriptions, ensure you have at least two Azure subscriptions or workspaces. These should not have Microsoft Defender for Cloud Plans enabled already.

Open a web browser and navigate to `https://portal.azure.com`.

How to do it...

To enable Microsoft Defender for Cloud Plans on multiple subscriptions at once, complete the following steps:

1. In the Azure portal, open **Microsoft Defender for Cloud**. You can open Microsoft Defender for Cloud in multiple ways: typing **Microsoft Defender for Cloud** in a search bar, clicking on a favorite link, or by going to **All Services**:

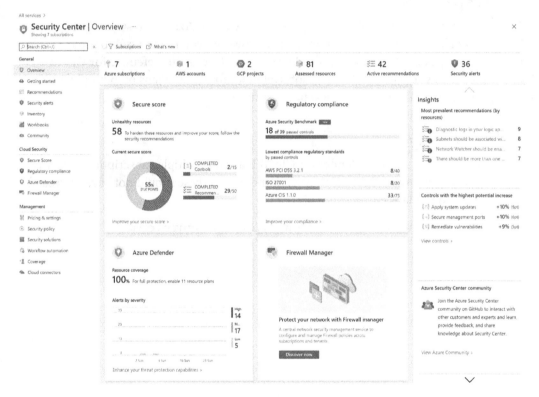

Figure 1.1 – Microsoft Defender for Cloud Overview page

2. On the **Microsoft Defender for Cloud** menu, on the left-hand side, select **Getting Started**. The **Getting Started** page will have three *tabs* or *pages* available: **Upgrade**, **Install Agents**, and **Get Started**. Click on **Upgrade** to display a list of available subscriptions and workspaces to enable on Microsoft Defender for Cloud Plans. The following screenshot shows this:

Figure 1.2 – Microsoft Defender for Cloud – Getting started page

3. Select all the subscriptions and workspaces you want to enable Microsoft Defender for Cloud Plans on and scroll to the end of the **Upgrade** page until the **Upgrade** button is visible. The **Upgrade** button is *gray* and will be disabled until you select at least one *subscription* or *workspace*, after which it will turn *blue*. Let's see what all of this looks like:

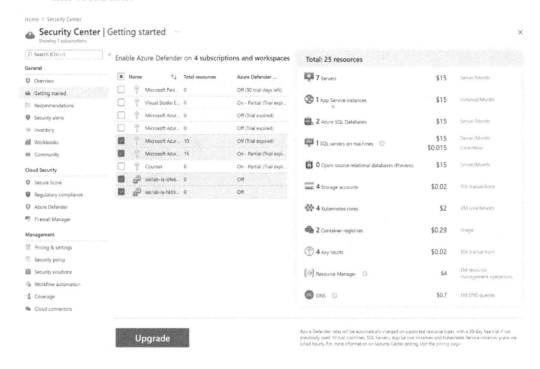

Figure 1.3 – Enabling Azure Subscriptions on Azure Subscriptions and/or Log Analytics Workspaces

4. Select **Upgrade** to enable Microsoft Defender for Cloud Plans on selected subscriptions and/or workspaces.

How it works...

As soon as you create an Azure Subscription, Microsoft Defender for Cloud gives you an overview of the resources that are monitored and assessed by Microsoft Defender for Cloud, as well as security recommendations for recognized resources.

To enable full coverage of Microsoft Defender for Cloud Plans on multiple Azure Subscriptions and workspaces, you can enable Microsoft Defender for Cloud Plans protection on more than one Subscription and Log Analytics workspace at once, reducing the risk of having unprotected resources and potential security issues.

Upgrading to and enabling full Microsoft Defender for Cloud Plans protection on multiple Azure Subscriptions and Log Analytics Workspaces applies to partially enabled Microsoft Defender for Cloud Plans as well.

Enabling an Microsoft Defender for Cloud Plans on an Azure Subscription

Microsoft Defender for Cloud covers two areas of cloud security: **Cloud Security Posture Management (CSPM)** and **Cloud Workload Protection (CWP)**. Microsoft Defender for Cloud Plans is Microsoft Defender for Cloud's integrated protection platform that protects Azure and hybrid resources. If you want to enable Microsoft Defender for Cloud Plans on a particular Azure Subscription and you want to control what Microsoft Defender for Cloud Plans features are enabled or disabled on an Azure Subscription, you need to enable Microsoft Defender for Cloud Plans, as described in this recipe.

There are multiple ways to enable Microsoft Defender for Cloud Plans on a subscription, and we will show more than one way here.

After completing this recipe, you will be able to enable Microsoft Defender for Cloud Plans and Microsoft Defender for Cloud Plans's protection features on an Azure Subscription.

Getting ready

Before you enable Microsoft Defender for Cloud Plans on an Azure Subscription, you must have at least one Azure Subscription. You should not have Microsoft Defender for Cloud Plans already enabled.

Open a web browser and navigate to `https://portal.azure.com`.

How to do it...

To enable Microsoft Defender for Cloud Plans and Microsoft Defender for Cloud Plans's protection capabilities on different workloads more granularly, complete the following steps:

1. In the Azure portal, open **Microsoft Defender for Cloud**. You can open Microsoft Defender for Cloud in multiple ways: by typing **Microsoft Defender for Cloud** in the search bar, clicking on a favorite link, or by going to **All Services**.

2. On the **Microsoft Defender for Cloud – Overview** page, from the left menu, select **Environment settings**, as shown in the following screenshot.

3. Expand Management groups until you see a desired Azure subscription

4. Select the subscription that you want to enable Microsoft Defender for Cloud Plans on. The **Settings – Defender Plans** page will open:

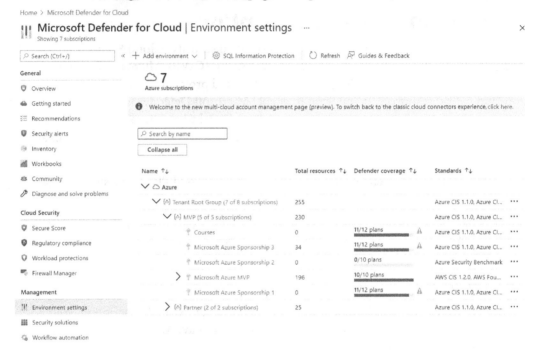

Figure 1.4 – Microsoft Defender for Cloud Environmental settings page – Selecting an Azure Subscription

5. While Microsoft Defender for Cloud Plans is disabled, all individual Microsoft Defender for Cloud Plans by resource types are grayed out and disabled. Click on **Microsoft Defender for Cloud Plans on** to enable Microsoft Defender for Cloud Plans, as shown in the following screenshot:

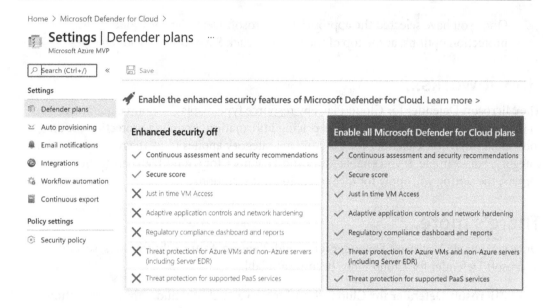

Figure 1.5 – Turning Microsoft Defender for Cloud Plans on

6. After you select **Enable all Microsoft Defender for Cloud Plans,** you can select Microsoft Defender for Cloud Plans by resource type individually. Alternatively, if you select an Azure Subscription that already has Microsoft Defender for Cloud Plans turned on partially, you can enable all Microsoft Defender for Cloud Plans by clicking on the **Enable all** button, as shown in the following screenshot. A button or control that has changed and its current setting is not saved will be *purple*; otherwise, it will be *blue*, as shown in the following screenshot:

Defender for Cloud plans will be enabled on 10 resources in this subscription

∧ Select Defender plan by resource type   Enable all

Microsoft Defender for	Resources	Pricing	Configuration	Plan
Servers	6 servers	$15/Server/Month ⓘ		On Off
App Service	1 instances	$15/Instance/Month ⓘ		On Off
Azure SQL Databases	0 servers	$15/Server/Month ⓘ		On Off
SQL servers on machines	0 servers	$15/Server/Month ⓘ $0.015/Core/Hour		On Off
Open-source relational databases	0 servers	$15/Server/Month ⓘ		On Off
Storage	3 storage accounts	$0.02/10k transactions ⓘ		On Off
Containers	0 container registries; 0 kubernetes ...	$7/VM core/Month ⓘ		On Off
Key Vault	3 key vaults	$0.02/10k transactions		On Off
Resource Manager		$4/1M resource management operatic		On Off
DNS		$0.7/1M DNS queries ⓘ		On Off

When you select Save, Microsoft Defender for Cloud's enhanced security features will be enabled on all the resource types you've selected. The first 30 days are free. For more information on Defender for Cloud pricing, visit the pricing page.

Figure 1.6 – Selecting Microsoft Defender for Cloud Plans by resource types

7. Once you have selected the appropriate Microsoft Defender for Cloud Plans
    protection options, at the top of the window, click **Save** to apply your changes.

How it works...

The Microsoft Defender for Cloud Plans by resource type displays resource quantities
in their respective categories, as well as pricing information. Enabling protection for an
individual Microsoft Defender for Cloud Plans category applies to *all* the resources in that
category. For example, if you enable Microsoft Defender for Cloud Plans protection for
servers, the setting will apply for *all* Servers in a subscription.

There's more...

Once you enable Microsoft Defender for Cloud Plans on Azure Subscriptions, several
Microsoft Defender for Cloud Plans become available:

- **Microsoft Defender for Cloud Plans for Servers**: This includes threat protection
  for Azure virtual machines and non-Azure servers, including server **Endpoint
  Detection and Response (EDR)**.

- **Microsoft Defender for Cloud Plans for App Service**.

- **Microsoft Defender for Cloud Plans for SQL Databases**: This applies to Azure
  SQL Database's single databases and elastic pools, Azure SQL Managed Instances,
  and Azure Synapse Analytics.

- **Microsoft Defender for Cloud Plans for SQL Servers on machines**: This applies to
  SQL on Azure virtual machines, SQL servers on-premises, and Azure Arc-enabled
  SQL servers.

- **Microsoft Defender for Cloud Plans for open source relational databases (at
  the time of writing, this feature is still in preview)**: This applies to non-Basic tier
  Azure Databases for a PostgreSQL single server, Azure Databases for a MySQL
  single server, and Azure Databases for a MariaDB single server.

- **Microsoft Defender for Cloud Plans for Storage**: This applies to blob containers,
  file shares, and DataLake Gen2.

- **Microsoft Defender for Cloud Plans for Containers,** which includes **Policy
  Add-on for Kubernetes, Azure Kubernetes Service profile** and **Azure
  Arc-Enabled Kubernetes extension** options.

- **Microsoft Defender for Cloud Plans for Key Vault**.

- **Microsoft Defender for Cloud Plans for Resource Manager**.

- **Microsoft Defender for Cloud Plans for DNS**: This applies to all Azure resources connected to the Azure default DNS.

Microsoft Defender for Cloud Plans's advanced protection capabilities include **Just-in-Time** virtual machine access, virtual machine vulnerability assessment, adaptive application control, adaptive network hardening, file integrity monitoring, SQL vulnerability assessment, container image scanning, **Internet of things** (**IoT**) security, regulatory compliance dashboards and reports, threat protection for supported PaaS services, and more.

Enabling an Microsoft Defender for Cloud Plans on a Log Analytics Workspace

Like enabling Microsoft Defender for Cloud Plans on an Azure Subscription, you can enable Microsoft Defender for Cloud Plans on a **Log Analytics Workspace**.

Getting ready

Before you enable Microsoft Defender for Cloud Plans on a Log Analytics Workspace, you must have at least one Log Analytics Workspace. You should not have Microsoft Defender for Cloud Plans already enabled.

Open a web browser and navigate to `https://portal.azure.com`.

How to do it...

To enable Microsoft Defender for Cloud Plans on a Log Analytics Workspace, complete the following steps:

1. In the Azure portal, open **Microsoft Defender for Cloud**. You can open Microsoft Defender for Cloud in multiple ways: by typing **Microsoft Defender for Cloud** in a search bar, clicking on a favorite link, or by going to **All Services**.

2. On the **Microsoft Defender for Cloud – Overview** page, from the left menu, select **Environmental settings**, as shown in the following screenshot.

3. Select the Log Analytics Workspace that you want to enable Microsoft Defender for Cloud Plans on. The **Settings – Defender Plans** page will open:

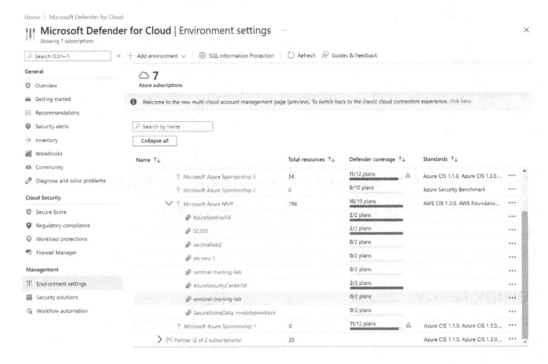

Figure 1.7 – Microsoft Defender for Cloud Getting started page – Selecting a Log Analytics Workspace

4. While Microsoft Defender for Cloud Plans is disabled, all individual Microsoft Defender for Cloud Plans by resource type will be grayed out and disabled. Click on **Microsoft Defender for Cloud Plans on**, as shown in the following screenshot, to enable Microsoft Defender for Cloud Plans:

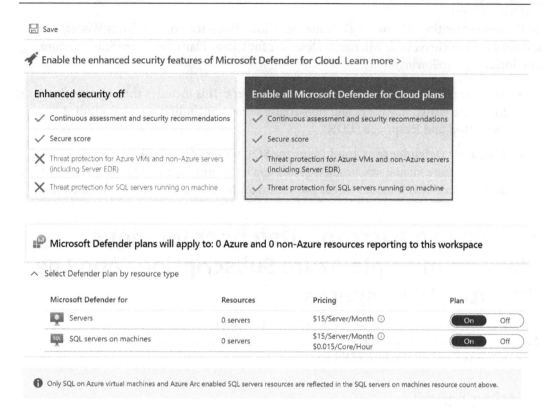

Figure 1.8 – Selecting and enabling Microsoft Defender for Cloud Plans on resource types reporting to a Log Analytics Workspace

5. Once you have selected Microsoft Defender for Cloud Plans's protection options, at the top of the window, click on **Save** to apply these changes.

How it works...

When you enable an Microsoft Defender for Cloud Plans for a resource type on a subscription, all the resource types in that category will be protected by Microsoft Defender for Cloud Plans. On the other hand, when you enable Microsoft Defender for Cloud Plans on a Log Analytics Workspace, Microsoft Defender for Cloud Plans will protect only resources reporting to that workspace.

At the time of writing, Microsoft Defender for Cloud Plans for Log Analytics Workspace are not as comprehensive as Microsoft Defender for Cloud Plans for Azure Subscriptions and include the following:

- **Microsoft Defender for Cloud Plans for Servers**: This includes threat protection for Azure virtual machines and non-Azure servers, including server **Endpoint Detection and Response (EDR)**.

- **Microsoft Defender for Cloud Plans for SQL servers on machines**: This applies to SQL on Azure virtual machines, SQL servers on-premises, and Azure Arc-enabled SQL servers. This includes threat protection for SQL servers running on machines.

Enabling an Microsoft Defender for Cloud Plans on multiple Azure Subscriptions and Log Analytics Workspaces

This recipe will show you how to enable Microsoft Defender for Cloud Plans and Microsoft Defender for Cloud Plans's protection capabilities on multiple Azure Subscriptions and Log Analytics Workspaces.

Getting ready

Before you enable Microsoft Defender for Cloud Plans on Azure Subscriptions and Log Analytics Workspaces, you must have at least one Azure subscription and at least one Log Analytics Workspace. You should not have Microsoft Defender for Cloud Plans already enabled on these subscriptions and workspaces.

Open a web browser and navigate to `https://portal.azure.com`.

How to do it...

To enable Microsoft Defender for Cloud Plans and Microsoft Defender for Cloud Plans's protection capabilities on multiple subscriptions and workspaces, complete the following steps:

1. In the Azure portal, open **Microsoft Defender for Cloud**. You can open Microsoft Defender for Cloud in multiple ways: by typing **Microsoft Defender for Cloud** in a search bar, clicking on a favorite link, or by going to **All Services**.

2. On the **Microsoft Defender for Cloud – Overview** page, from the left menu, select **Getting Started**, as shown in the following screenshot. A list of available Azure Subscriptions should be displayed. If not, select the **Upgrade** tab at the top of the page:

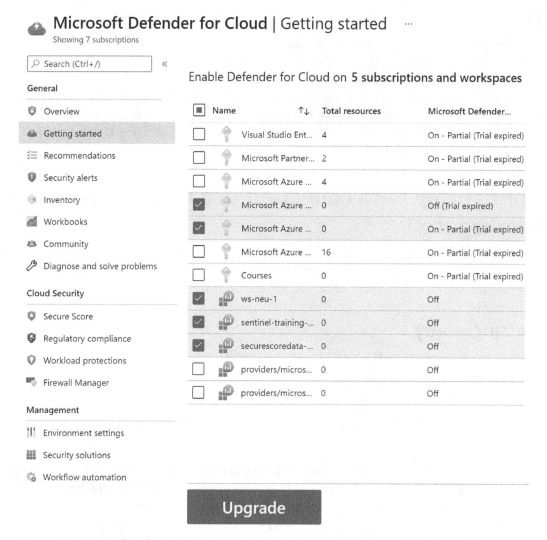

Figure 1.9 – Microsoft Defender for Cloud Getting started page – Selecting multiple Azure Subscriptions and Log Analytics Workspaces

3. In the middle of the page, under **Select subscriptions and workspaces to enable Microsoft Defender for Cloud Plans on**, select the checkboxes next to the *Azure Subscriptions* and *Log Analytics Workspaces* you want to enable Microsoft Defender for Cloud Plans on.

4. At the bottom of the page, click on **Upgrade**.

How it works...

After enabling Microsoft Defender for Cloud Plans on multiple Azure Subscriptions and Log Analytics Workspaces, you have the option to install Log Analytics Agent on recently enabled subscriptions and workspaces at the same time by selecting the **Install** button. Similar to the previous recipe, enabling protection for individual Microsoft Defender for Cloud Plans categories applies to *all* the resources in that category. For example, if you enable Microsoft Defender for Cloud Plans protection for servers, this setting will apply to *all* the servers in a subscription.

Configuring data collection in a Log Analytics Workspace

You can **configure data collection** tiers in a Log Analytics Workspace. This will affect the number and type of events stored in a Log Analytics Workspace. The data that's stored in a workspace allows you to search, audit, and investigate stored events.

Getting ready

Open a web browser and navigate to `https://portal.azure.com`.

How to do it...

To configure the level of data you wish to store in a Log Analytics Workspace, complete the following steps:

1. In the Azure portal, open **Microsoft Defender for Cloud**. You can open Microsoft Defender for Cloud in multiple ways: by typing **Microsoft Defender for Cloud** in a search bar, clicking on a favorite link, or by going to **All Services**.

2. On the **Microsoft Defender for Cloud – Overview** page, from the left menu, select **Environmental settings**.

3. Select the Log Analytics Workspace that you want to configure the level of data to store for. The **Settings – Defender Plans** page will open.

4. If Microsoft Defender for Cloud Plans, if the Log Analytics Workspace is turned *off*, you must turn Microsoft Defender for Cloud Plans *on*.

    Please refer to the *Enabling Microsoft Defender for Cloud Plans on a Log Analytics Workspace* recipe for a complete procedure regarding how to enable Microsoft Defender for Cloud Plans on a Log Analytics Workspace.

5. From the left menu, select **Data collection**, as shown in the following screenshot:

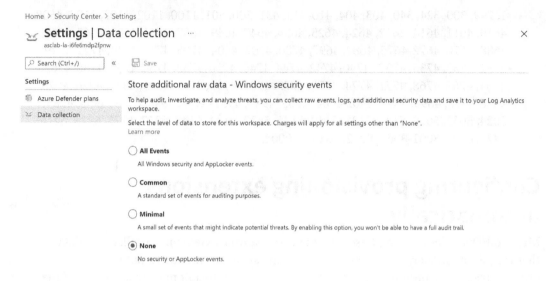

Figure 1.10 – Selecting the level of data to store for a selected Log Analytics Workspace

6. By default, data collection is turned off. To choose the level of data to store in the selected workspace, select either **Minimal**, **Common**, or **All Events**.

7. Click on **Save** to apply these changes.

How it works...

A **Log Analytics Agent** collects events required for Microsoft Defender for Cloud protection, regardless of the level of data stored in the Log Analytics Workspace. You can choose to collect various types of events in a Log Analytics Workspace, such as logs, raw events, and security data, by selecting the level of data to store for a workspace. Setting the option to something other than None enables you to search for, audit, and investigate collected events in Microsoft Defender for Cloud.

The **Minimal** setting includes collecting the following events:

*   1102, 4624, 4625, 4657, 4663, 4688, 4700, 4702, 4719, 4720, 4722, 4723, 4724, 4727, 4728, 4732, 4735, 4737, 4739, 4740, 4754, 4755, 4756, 4767, 4799, 4825, 4946, 4948, 4956, 5024, 5033, 8001, 8002, 8003, 8004, 8005, 8006, 8007, 8222.

The **Common** setting includes collecting the following events:

- 1, 299, 300, 324, 340, 403, 404, 410-413, 431, 500, 501, 1100, 1102, 1107, 1108, 4608, 4610, 4611, 4614, 4622, 4624, 4625, 4634, 4647-4649, 4657, 4661-4663, 4665-4667, 4688, 4670, 4672-4675, 4689, 4697, 4700, 4702, 4704, 4705, 4716-4720, 4722-4729, 4732, 4733, 4735, 4737- 4740, 4742, 4744-4746, 4750-4752, 4754-4757, 4760-4762, 4764, 4767, 4768, 4771, 4774, 4778, 4779, 4781, 4793, 4797-4803, 4825, 4826, 4870, 4886-4888, 4893, 4898, 4902 4904, 4905, 4907, 4931-4933, 4946, 4948, 4956, 4985, 5024, 5033, 5059, 5136, 5137, 5140, 5145, 5632, 6144, 6145, 6272, 6273, 6278, 6416, 6423, 6424, 8001-8007, 8222, 26401, 30004.

Configuring provisioning extensions automatically

Microsoft Defender for Cloud uses an **agent** or **resource extension** to collect the data that's required for analysis. You can choose to install agents manually, but for the best protection and less administrative burden, you may wish to automate installations of the monitoring agent and its extensions.

Getting ready

Open a web browser and navigate to `https://portal.azure.com`. Microsoft Defender for Cloud Plans must be enabled on the Azure Subscription you are configuring.

How to do it...

To ensure you have the required resource extensions or agents installed automatically, complete the following steps:

1. In the Azure portal, open **Microsoft Defender for Cloud**. You can open Microsoft Defender for Cloud in multiple ways: by typing **Microsoft Defender for Cloud** in a search bar, clicking on a favorite link, or by going to **All Services**.

2. On the **Microsoft Defender for Cloud – Overview** page, from the left menu, select **Environmental settings**.

3. Select the Azure Subscription that you want to configure automatic provisioning of agents and extensions on. The **Settings – Defender Plans** page will open. Next, we'll be working with what's shown in the following screenshot:

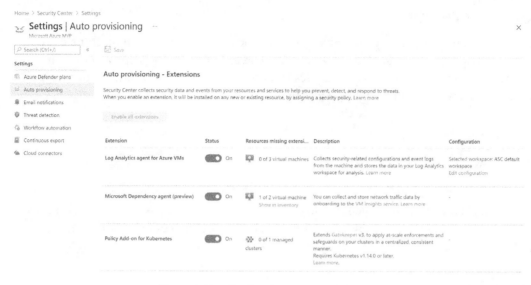

Figure 1.11 – Configuring auto provisioning

4. From the left menu, select **Auto provisioning**. The **Auto provisioning – Extensions** page will open.

5. Click on the **Status** button for the respective extension to enable an extension or to turn auto-provisioning *on*.

How it works...

When you enable an extension – that is, turn auto-provisioning on – any new or existing resources will have an extension or agent automatically installed. This can be achieved by assigning the proper **Deploy if not exist** policy that applies to all present and future resources.

Enabling a Log Analytics agent for Azure VMs manually in the Log Analytics Workspace settings

Let's say you want to enable Log Analytics Agent on an Azure virtual machine manually. This recipe will explain how to perform such an installation using the Log Analytics Workspace **blade settings**.

Getting ready

Assuming auto-provisioning is disabled and that the target Azure virtual machine does not have Log Analytics Agent already installed, you can perform the steps described in this recipe. You must have a Log Analytics Workspace provisioned to conduct this recipe. Open a web browser and navigate to `https://portal.azure.com`.

How to do it...

The following steps are to be performed:

1. In the Azure portal, open the **Log Analytics Workspaces** blade where you want to enable Log Analytics Agent manually. You can open the **Log Analytics workspaces** blade in multiple ways: by typing **Log Analytics workspaces** in a search bar, clicking on a favorite link, or by going to **All Services**:

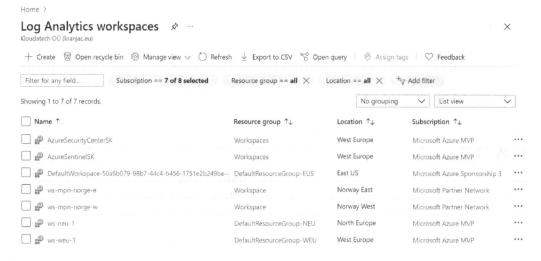

Figure 1.12 – Selecting a Log Analytics Workspace

2. Click on a workspace to open the **Workspace** blade.

3. From the left menu, under **Workspace Data Sources**, select **Virtual Machines**, as shown in the following screenshot:

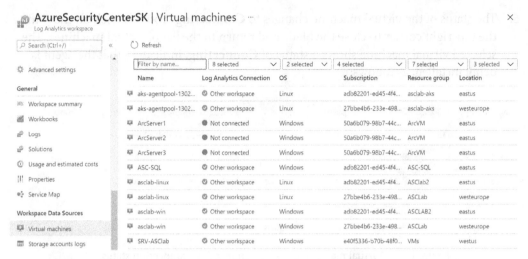

Figure 1.13 – Selecting a virtual machine

4. The list of virtual machines shows their respective connection status in the **Log Analytics Connection** column: *Not connected*, *Connecting*, *This workspace*, or *Other workspace*. Select a virtual machine that you want to install the Log Analytics Agent on manually and connect it to the selected Log Analytics Workspace.

5. Click on **Connect** to begin installing the agent on the virtual machine and connect it to the workspace, as shown in the following screenshot:

Figure 1.14 – Connecting a virtual machine to a workspace

6. The status of the virtual machine changes to **Connecting**. You can click on the *X* in the top-right corner to close the blade and return to the list of virtual machines. The selected virtual machine shows a status of **Connecting**, indicating that the agent is being installed, as follows:

🖵 aks-agentpool-1302...	✅ Other workspace	Linux	adb82201-ed45-4f4...	asclab-aks	eastus
🖵 aks-agentpool-1302...	✅ Other workspace	Linux	27bbe4b6-233e-498...	asclab-aks	westeurope
🖵 ArcServer1	⊕ Connecting	Windows	50a6b079-98b7-44c...	ArcVM	eastus
🖵 ArcServer2	⬤ Not connected	Windows	50a6b079-98b7-44c...	ArcVM	eastus
🖵 ArcServer3	⬤ Not connected	Windows	50a6b079-98b7-44c...	ArcVM	eastus
🖵 ASC-SQL	✅ Other workspace	Windows	adb82201-ed45-4f4...	ASC-SQL	eastus

Figure 1.15 – Virtual machine showing "Connecting" connection status

7. After a few moments, check the installation status. When the installation process is completed and the agent has been installed on the virtual machine, its status will change to **This workspace**, as follows:

🖵 aks-agentpool-1302...	✅ Other workspace	Linux	adb82201-ed45-4f4...	asclab-aks	eastus
🖵 aks-agentpool-1302...	✅ Other workspace	Linux	27bbe4b6-233e-498...	asclab-aks	westeurope
🖵 ArcServer1	✅ This workspace	Windows	50a6b079-98b7-44c...	ArcVM	eastus
🖵 ArcServer2	⬤ Not connected	Windows	50a6b079-98b7-44c...	ArcVM	eastus
🖵 ArcServer3	⬤ Not connected	Windows	50a6b079-98b7-44c...	ArcVM	eastus
🖵 ASC-SQL	✅ Other workspace	Windows	adb82201-ed45-4f4...	ASC-SQL	eastus

Figure 1.16 – Virtual machine showing "This Workspace" connection status

How it works...

There are several reasons why you would want to install Log Analytics Agent manually – perhaps you wanted to have greater control over software installations on your virtual machines and you disabled auto-provisioning. In that case, Microsoft Defender for Cloud does not install agents automatically and you must perform a manual installation on Azure virtual machines.

In this example, a Log Analytics Workspace has shown the list of virtual machines and their workspace connection statuses. From the list of available virtual machines, it is then easy to connect them to the selected workspace.

Enabling a Log Analytics agent for Azure VMs manually in the Virtual Machine settings

Let's say you want to enable Log Analytics Agent on an Azure virtual machine manually. This recipe will explain how to perform such an installation using the **Virtual Machine** blade settings.

Getting ready

Assuming auto-provisioning is disabled and the target Azure virtual machine does not have Log Analytics Agent already installed, you can perform the steps described in this recipe. You must have a Log Analytics Workspace provisioned if you do not have one. Open a web browser and navigate to `https://portal.azure.com`.

How to do it...

1. In the Azure portal, open the **Virtual Machine** blade where you want to enable Log Analytics Agent manually. You can open the **Virtual Machine** blade in multiple ways: by typing **Virtual Machine** in a search bar, clicking on a favorite link, or by going to **All Services**.

2.   From the left menu, under **Monitoring**, select **Logs**:

Figure 1.17 – Enabling monitoring agent on the Virtual Machine Logs blade

3.   On the **Logs** page, click on the **Enable** button. This button will briefly change to **Validating**, after which the following menu will become available:

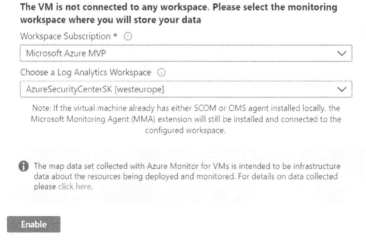

Figure 1.18 – Choosing a workspace subscription and a Log Analytics Workspace

4. From the pull-down menus shown in the preceding screenshot, choose an *Azure Subscription* and a *Log Analytics Workspace* you want to connect the virtual machine to.

5. Click **Enable** to begin installing Log Analytics Agent.

How it works...

There are several reasons why you would want to install Log Analytics Agent manually – perhaps you wanted to have greater control over software installations on your virtual machines and you disabled auto-provisioning. In that case, Microsoft Defender for Cloud does not install agents automatically and you must perform a manual installation on Azure virtual machines.

In this example, once monitoring has been enabled on a virtual machine, you can choose a workspace to connect to from a list of available workspaces.

There's more...

Optionally, you can check if Log Analytics Agent was installed successfully in several ways. One way would be to check the list of connected virtual machines on the Log Analytics Workspace's list of connected virtual machines, as follows:

Filter by name...	8 selected ∨	2 selected ∨	4 selected ∨	7 selected ∨	3 selected ∨
Name	Log Analytics Connection	OS	Subscription	Resource group	Location
aks-agentpool-13023423-0	⊘ Other workspace	Linux	adb82201-ed45-4f4d-b9a...	asclab-aks	eastus
aks-agentpool-13023423-0	⊘ Other workspace	Linux	27bbe4b6-233e-4985-891...	asclab-aks	westeurope
ArcServer1	⊘ This workspace	Windows	50a6b079-98b7-44c4-b45...	ArcVM	eastus
ArcServer2	⊘ This workspace	Windows	50a6b079-98b7-44c4-b45...	ArcVM	eastus
ArcServer3	● Not connected	Windows	50a6b079-98b7-44c4-b45...	ArcVM	eastus
ASC-SQL	⊘ Other workspace	Windows	adb82201-ed45-4f4d-b9a...	ASC-SQL	eastus
asclab-linux	⊘ Other workspace	Linux	adb82201-ed45-4f4d-b9a...	ASClab2	eastus

Figure 1.19 – Checking the virtual machine workspace's connection status

You can check the **Logs** blade on the connected virtual machine pages. If you installed this successfully, you should be able to see the **Workspace** query blade:

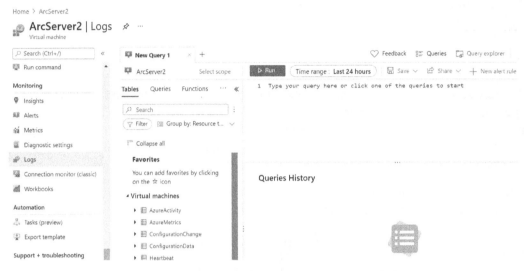

Figure 1.20 – Virtual machine Log analytics Query blade

Moreover, on a virtual machine's **Overview** page, you can check the extensions' installation status as well. This list will show the names of the agents that have been installed. In this case, both *Windows Dependency Agent* and *Microsoft Monitoring Agent* have been successfully installed, as follows:

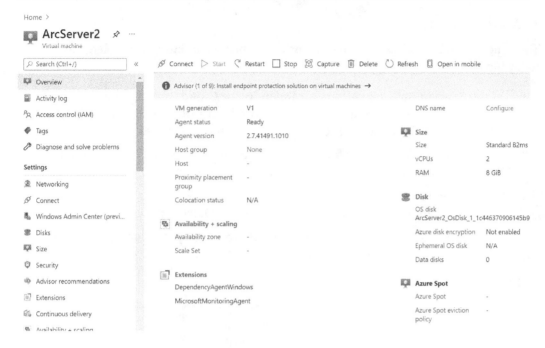

Figure 1.21 – Virtual Machine Overview blade showing Extensions

Configuring a Log Analytics agent for Azure VMs extension deployment

Similar to configuring the level of data that's collected by Log Analytics Workspace, you can configure data collection tiers at the Azure Subscription level, and it will affect the number and type of events stored for the Azure Subscription as well. The data that's collected by Microsoft Defender for Cloud is stored in Log Analytics Workspace so that you can search for, audit, and investigate stored events.

Getting ready

Open a web browser and navigate to `https://portal.azure.com`.

How to do it...

To configure the Log Analytics Agent extension's deployment settings, complete the following steps:

1. In the Azure portal, open **Microsoft Defender for Cloud**. You can open Microsoft Defender for Cloud in multiple ways: by typing **Microsoft Defender for Cloud** in a search bar, clicking on a favorite link, or by going to **All Services**.

2. On the **Microsoft Defender for Cloud – Overview** page, from the left menu, select **Environmental settings**.

3. Select the Azure Subscription that you want to configure the Log Analytics Agent extension's deployment settings for. The **Settings – Defender Plans** page will open.

4. From the left menu, select **Auto provisioning**. The **Auto provisioning – Extension** page will open, as follows:

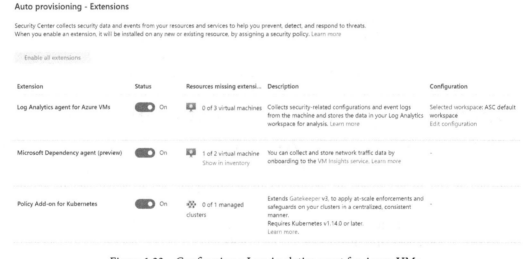

Figure 1.22 – Configuring a Log Analytics agent for Azure VMs

5. On the right-hand side of the page, under **Configuration** for **Log Analytics agent for Azure VMs**, select **Edit configuration**. On the right-hand side, the **Extension deployment configuration** blade will open, as follows:

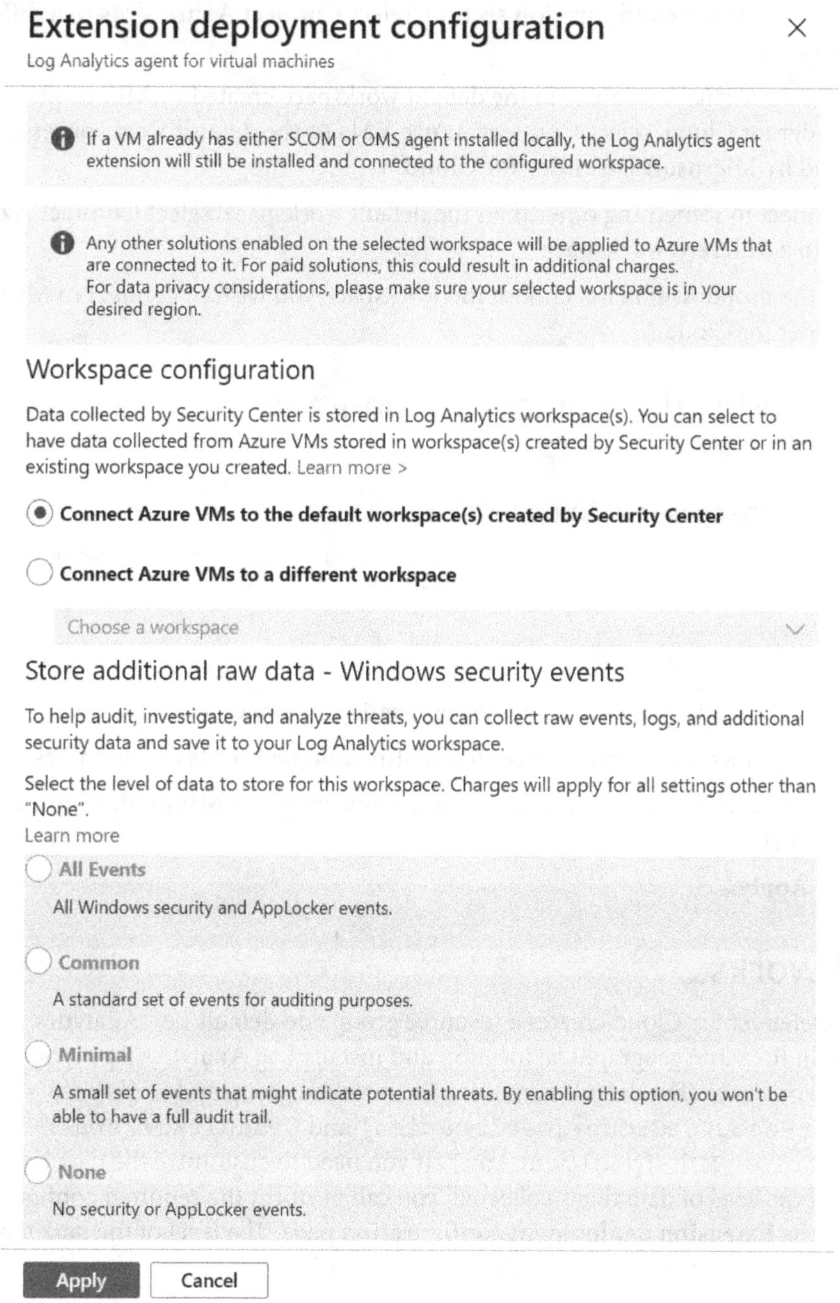

Extension deployment configuration   ✕

Log Analytics agent for virtual machines

> ℹ If a VM already has either SCOM or OMS agent installed locally, the Log Analytics agent
> extension will still be installed and connected to the configured workspace.

> ℹ Any other solutions enabled on the selected workspace will be applied to Azure VMs that
> are connected to it. For paid solutions, this could result in additional charges.
> For data privacy considerations, please make sure your selected workspace is in your
> desired region.

Workspace configuration

Data collected by Security Center is stored in Log Analytics workspace(s). You can select to
have data collected from Azure VMs stored in workspace(s) created by Security Center or in an
existing workspace you created. Learn more >

◉ **Connect Azure VMs to the default workspace(s) created by Security Center**

◯ **Connect Azure VMs to a different workspace**

Choose a workspace   ⌄

Store additional raw data - Windows security events

To help audit, investigate, and analyze threats, you can collect raw events, logs, and additional
security data and save it to your Log Analytics workspace.

Select the level of data to store for this workspace. Charges will apply for all settings other than
"None".
Learn more

◯ All Events

   All Windows security and AppLocker events.

◯ Common

   A standard set of events for auditing purposes.

◯ Minimal

   A small set of events that might indicate potential threats. By enabling this option, you won't be
   able to have a full audit trail.

◯ None

   No security or AppLocker events.

Apply   Cancel

Figure 1.23 – Configuring a Log Analytics agent for virtual machines deployment

6. In the **Workspace configuration** section, select **Connect Azure VMs to a different workspace**.

7. To connect virtual machines to the default workspace created by Microsoft Defender for Cloud, select **Connect Azure VMs to the default workspace(s) created by Microsoft Defender for Cloud**.

8. To connect to something other than the default workspace, select **Connect Azure VMs to a different workspace**.

9. From the dropdown menu, choose the workspace you wish to connect to Microsoft Defender for Cloud.:

Figure 1.24 – Applying changes to the extension deployment configuration

10. Select if you want to apply changes to **Existing and new VMs** or **New VMs only**.

11. Select the level of data to store for the selected workspace: **Minimal**, **Common**, or **All Events**.

12. Select **Apply**.

How it works...

Microsoft Defender for Cloud creates a resource group and default Log Analytics Workspace in the same geographical location and installs Log Analytics Agent on the default workspace. The default names for the resource group and workspace are DefaultResourceGroup-[geolocation] and DefaultWorkspace-[subscription-ID]-[geolocation]. If you need to customize the data collection location and the level of data that's collected, you can perform the required configuration changes on the **Extension deployment configuration** page. The level of the data that's collected depends on the setting you selected in the configuration.

Configuring email notifications

As a default behavior, Azure Subscription owners receive emails every time a high severity alert is activated for a subscription. To configure additional email recipients and the level of alerts to be notified, you have the option to configure the **Email notifications** settings for an Azure Subscription.

Getting ready

Open a web browser and navigate to `https://portal.azure.com`.

How to do it...

To configure email notifications for an Azure Subscription, complete the following steps:

1. In the Azure portal, open **Microsoft Defender for Cloud**. You can open Microsoft Defender for Cloud in multiple ways: by typing **Microsoft Defender for Cloud** in a search bar, clicking on a favorite link, or by going to **All Services**.

2. On the **Microsoft Defender for Cloud – Overview** page, from the left menu, select **Environmental settings**.

3. Select the Azure Subscription that you want to configure the **Email notifications** settings for. The **Settings – Defender Plans** page will open.

4. From the left menu, select **Email notifications**. The **Settings – Email notifications** page will open, as follows:

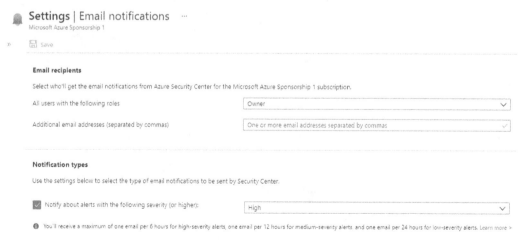

Figure 1.25 – Configuring email notifications

5. From the **All users with the following roles** dropdown menu, select users with a specific Azure role to be notified of alert events, as follows:

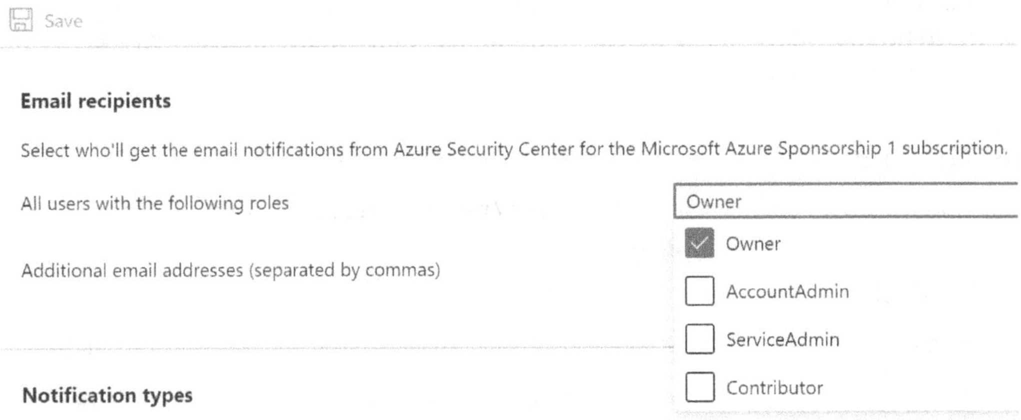

Figure 1.26 – Configuring email recipients

6. In the **Additional email addresses** field, enter one or more email addresses separated by commas to add additional users as alerts recipients, as shown in the following screenshot:

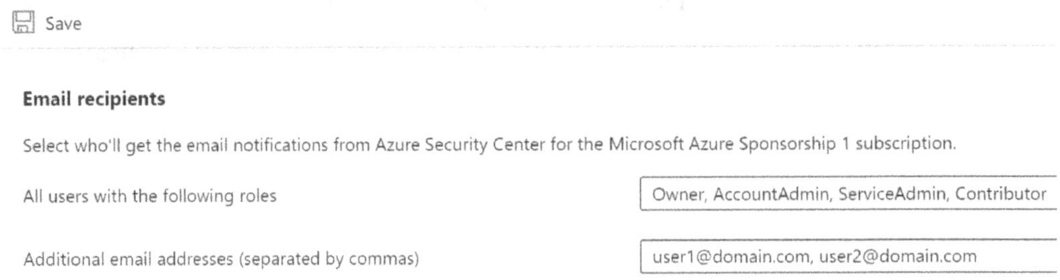

Figure 1.27 – Configuring additional email addresses

7. To turn email notifications *on* or *off*, *select* or *deselect* the checkbox for Notify about alerts with the following severity (or higher), and to set the level or the notification types of emails to be sent by Microsoft Defender for Cloud, select the adjacent dropdown menu and choose High, Medium, or Low:

Notification types

Use the settings below to select the type of email notifications to be sent by Security Center.

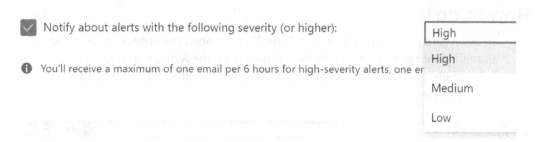

Figure 1.28 – Configuring notification types

8. Select **Save** to confirm and apply your changes.

How it works...

Azure subscription owners receive alert email notifications by default. To include additional recipients or the control frequency and type of alert notifications users are notified of, you need to configure further settings. A user with the *Owner*, *Contributor*, *Account Administrator*, and *Service Administrator* roles can receive alert notifications.

Depending on a configured level of alert notifications, Microsoft Defender for Cloud sends a maximum of one email per the following:

- 24 hours for low severity events (or one email per day)
- 12 hours for medium severity alerts (or two emails per day)
- 6 hours for high severity alerts (or four emails per day)

Assigning Microsoft Defender for Cloud permissions

Like other resources and services in Azure, **role-based access control** (**RBAC**) roles are the way to control rights and allow actions on Microsoft Defender for Cloud. In this recipe, you will assign appropriate RBAC roles to Microsoft Defender for Cloud for an Azure user.

Getting ready

Open a web browser and navigate to `https://portal.azure.com`.

How to do it...

1. In the Azure portal, open **Subscriptions**. You can open the **Subscriptions** blade in multiple ways: by selecting **Subscriptions** from the Azure portal main page, typing **Subscriptions** in a search bar, clicking on a favorite link, or by going to **All Services**:

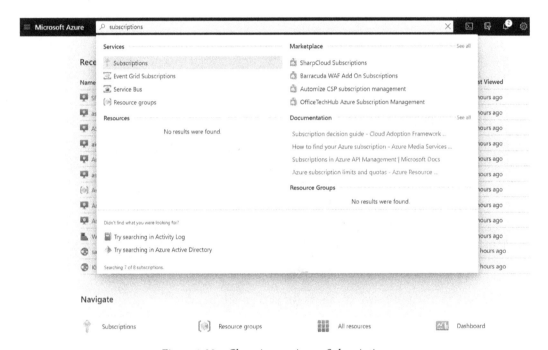

Figure 1.29 – Choosing an Azure Subscription

2. On the **Subscriptions** blade, select the subscription you want to configure Microsoft Defender for Cloud permissions on.

3. On the **Azure subscription** blade, from the left-hand side menu, select **Access Control (IAM)**. Then, from the right-hand side of the **Azure subscription** blade, at the top, select **Role assignments**:

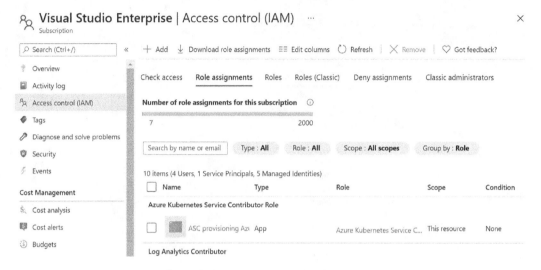

Figure 1.30 – Selecting a Role assignment

4. Select **Add**:

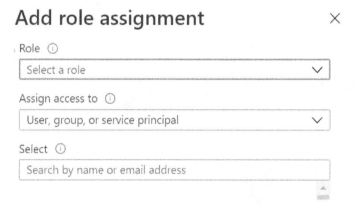

Figure 1.31 – Add role assignment blade

5. On the **Add role assignment** page, under the **Role** menu, select the **Security Admin** role. Under **Assign access to**, select **User, group, service principal**. Under **Select**, select one or more members you want to assign a role to. Select **Save**:

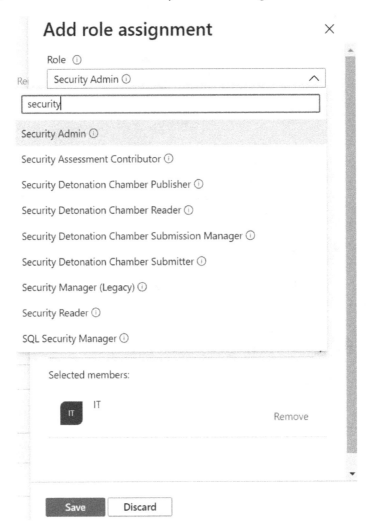

Figure 1.32 – Choosing a role

6. When the **Access control (IAM)** blade is open, repeat *Steps 4* and *5* to add the **Security Reader** role.

How it works...

There are two Microsoft Defender for Cloud roles available called **Security Reader** and **Security Admin** that allow specific MDC actions to be performed. In addition to these two roles, you can perform certain Azure Security Actions when you are assigned the role of **Owner**, **Contributor**, or **Reader**.

The following table shows the allowed actions and roles in Microsoft Defender for Cloud:

Action	Security Reader/Reader	Security Admin	Resource Group Contributor/Resource Group Owner	Subscription Contributor	Subscription Owner
Edit security policy	-	✔	-	-	✔
Add/assign initiatives (including regulatory compliance standards)	-	-	-	-	✔
Enable/disable Microsoft Defender for Cloud Plans	-	✔	-	-	✔
Enable/disable auto-provisioning	-	✔	-	✔	✔
Apply security recommendations for a resource (and use Fix)	-	-	✔	✔	✔
Dismiss alerts	-	✔	-	✔	✔
View alerts and recommendations	✔	✔	✔	✔	✔

Table 1.1 – Allowed actions and roles in Microsoft Defender for Cloud

To be able to use **Workflow Automation** in Microsoft Defender for Cloud, you must have appropriate permissions on **Logic Apps**, which are used in automation scenarios:

- `Microsoft.Security/automations/write` on the resource group where the automation workflow will be created.

- **Logic App Contributor** on a Logic App you are creating.

Onboarding Microsoft Defender for Cloud using PowerShell

You can onboard Microsoft Defender for Cloud and perform the initial configuration steps using **PowerShell**. These steps include setting the Microsoft Defender for Cloud coverage level, configuring Log Analytics Workspace, and installing an agent.

Getting ready

Before executing Microsoft Defender for Cloud PowerShell commands, you must perform some initial steps. Run PowerShell with **administrative privileges**. The latest *Az* modules need to be installed, as follows:

```
Install-Module -Name Az -AllowClobber -Scope CurrentUser
```

Ensure that an execution policy has been set:

```
Set-ExecutionPolicy -ExecutionPolicy AllSigned
```

Ensure that the Az.Security module has been installed:

```
Install-Module -Name Az.Security -Force
```

How to do it...

To onboard Microsoft Defender for Cloud using PowerShell, complete the following steps:

1. Register your subscriptions to the **Microsoft Defender for Cloud Resource Provider**:

    ```
    Set-AzContext -Subscription "subscription_ID"
    Register-AzResourceProvider '
    -ProviderNamespace 'Microsoft.Security'.
    ```

2. Agents will have to report to the Log Analytics Workspace you are configuring, which the agents will report. You must have a Log Analytics Workspace that you have already created that the subscription's VMs will report to. You can define multiple subscriptions to report to the same workspace. If this is not defined, the default workspace will be used:

    ```
    Set-AzSecurityWorkspaceSetting -Name "default" '
    -Scope "/subscriptions/<subscription_ID> " '
    ```

```
-WorkspaceId "/subscriptions/<subscription_ID> /
resourceGroups/<ResourceGroupName>/providers/Microsoft.
OperationalInsights/workspaces/<WorkspaceName>"
```

3. Auto-provision the installation of Log Analytics Agent on your Azure VMs:

```
Set-AzContext -Subscription "<subscription_ID>"
Set-AzSecurityAutoProvisioningSetting '
-Name "default" -EnableAutoProvision
```

4. Assign the default **Microsoft Defender for Cloud** policy initiative:

```
Register-AzResourceProvider '
-ProviderNamespace 'Microsoft.PolicyInsights'
$Policy = Get-AzPolicySetDefinition | where {$_.
Properties.displayName -EQ 'Azure Security Benchmark'}
New-AzPolicyAssignment '
-Name 'ASC Default <subscription_ID>' '
-DisplayName 'Microsoft Defender for Cloud Default
<subscription ID>' '
-PolicySetDefinition $Policy '
-Scope '/subscriptions/<subscription_ID>'
```

Replace <subscription_ID> with the ID of an Azure Subscription,
<WorkspaceName> with the Log Analytics **Workspace ID**, and
<ResourceGroupName> with a name of a resource group.

How it works...

To onboard Microsoft Defender for Cloud using PowerShell, you must use the
Az.Security PowerShell module. First, you must register your subscriptions to the
Microsoft Defender for Cloud resource provider, then configure the Log Analytics
Workspace that Log Analytics Agent will report to. Finally, you must ensure auto-
provisioning for Log Analytics Agent has been configured on your Azure virtual machines
and assign the default Azure Security policy initiative. Optionally, you can set Microsoft
Defender for Cloud Plans to enabled.

There's more...

Optionally, you can enable Microsoft Defender for Cloud Plans using the following command:

```
Set-AzContext -Subscription "<subscription ID>"
```

You can also use the following command to do the same:

```
Set-AzSecurityPricing -Name "default" -PricingTier "Standard"
```

Enabling Microsoft Defender for Cloud integration with other Microsoft security services

Microsoft Defender for Cloud extends its protection capabilities beyond Microsoft Defender for Cloud Plans security. To configure additional threat protection capacity in Microsoft Defender for Cloud, Microsoft Defender for Cloud Plans must be enabled.

Getting ready

Open a web browser and navigate to `https://portal.azure.com`.

How to do it...

To configure and enable Microsoft Defender for Cloud integration with **Microsoft Cloud App Security** and Microsoft Defender for Endpoint, complete the following steps:

1. In the Azure portal, open **Microsoft Defender for Cloud**. You can open Microsoft Defender for Cloud in multiple ways: by typing **Microsoft Defender for Cloud** in a search bar, clicking on a favorite link, or by going to **All Services**.

2. On the **Microsoft Defender for Cloud – Overview** page, from the left menu, select **Environmental settings**.

3. Select the Azure Subscription that you want to configure integration with security services for. The **Settings – Defender Plans** page will open.

4. From the left menu, select **Integrations**. The **Enable integrations** page will open, as follows:

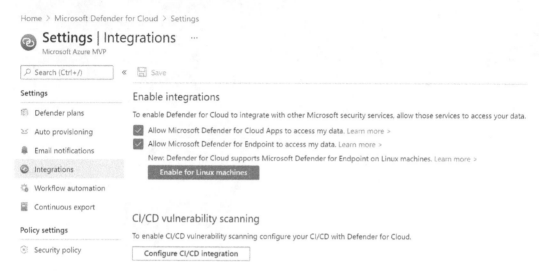

Figure 1.33 – Enabling Cloud App Security and Microsoft Defender for endpoint integration

5. To enable Microsoft Defender for Cloud integration with `Microsoft Cloud App Security` and `Microsoft Defender for Endpoint`, select their respective checkboxes.

6. Click **Save** to apply your changes.

How it works...

Microsoft Defender for Cloud uses specific packet headers based on **Internet Protocol Flow Information Export (IPFIX)** data to analyze network layer data that enables network-related alerts to be generated. If you have a **Microsoft Cloud App Security (MCAS)** license, some of the network-related alerts are operated by MCAS and these alerts are enabled by default. Moreover, by integrating Microsoft Defender for Endpoint with Microsoft Defender for Cloud, additional endpoint protection capabilities become available: **automated onboarding** for all Windows servers monitored by Microsoft Defender for Cloud and **centralized** monitoring of Microsoft Defender for Endpoint alerts in Microsoft Defender for Cloud.

2
Multi-Cloud Connectivity

In this chapter, you will learn how to connect your hybrid and multi-cloud computers to Azure and enable **Microsoft Defender for Cloud** to monitor the security posture of these connected resources.

Connecting your computing resources to Azure provides the convenience of monitoring their security posture through a single pane of glass, seeing recommendations, and performing security-related actions from a central place.

Connecting non-Azure computers to Azure involves several steps and, in this chapter, more complex actions will be divided into multiple recipes, each covering individual steps toward a bigger objective.

We will cover the following recipes in this chapter:

- Connecting non-Azure virtual machines using Azure Arc
- Connecting non-Azure virtual machines using Microsoft Defender for Cloud portal pages
- Setting up **Amazon Web Services Config** and **Amazon Web Services Security Hub**
- Creating an **Identity and Access Management (IAM) Amazon Web Services (AWS)** role for Microsoft Defender for Cloud

- Connecting Amazon Web Services to Microsoft Defender for Cloud
- Configuring **GCP Security Command Center** and enabling **GCP Security Command Center API**
- Creating a **GCP service account** and connecting **GCP** to Microsoft Defender for Cloud

Technical requirements

To complete the recipes in this chapter, the following are required:

- An Azure subscription
- An **Amazon Web Services (AWS)** account
- A **Google Cloud Platform (GCP)** account
- Azure PowerShell
- A web browser, preferably Microsoft Edge

The code samples for this chapter can be found at `https://github.com/PacktPublishing/Microsoft-Defender-for-Cloud-Cookbook`.

Connecting non-Azure virtual machines using Azure Arc

Before Microsoft Defender for Cloud can monitor your security posture and display security recommendations of your non-Azure computers, you must connect them to Azure. This recipe will show you how to connect a non-Azure server to Azure using Azure Arc.

Getting ready

Before you start connecting servers to Azure using Azure Arc, you must have administrative permissions on a target server to install and configure it.

Open a web browser and navigate to `https://portal.azure.com`.

How to do it...

To enable Microsoft Defender for Cloud Plans on multiple subscriptions at once, complete the following steps:

1. In the Azure portal, open **Azure Arc**. You can open Azure Arc in multiple ways: by typing **Azure Arc** in the search bar, by going to **All Services,** or by clicking on the respective link in **Favorites**.

2. On the left menu, select **Servers**:

Figure 2.1 – Selecting Azure Arc | Servers

3. At the top of the **Azure Arc | Servers** blade, select **+ Add**:

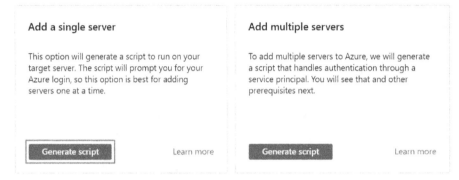

Figure 2.2 – Add servers with Azure Arc

4. In **Add single server box**, select **Generate script**.

5. On the ❶ **Prerequisites** blade, read the information provided and click **Next: Resource details** at the bottom of the blade.

6. On the ❷ **Resource details** blade, select the subscription and resource group that will contain the server you are adding and that will be managed in Azure. In addition, select the Azure region where the metadata for resources will be created, and the operating system that the server is running. If the server you are adding is unable to communicate directly to the internet and Azure data centers, specify the URL of the proxy server that the non-Azure server can use. At the bottom of the page, click **Next: Tags** to proceed to the next step:

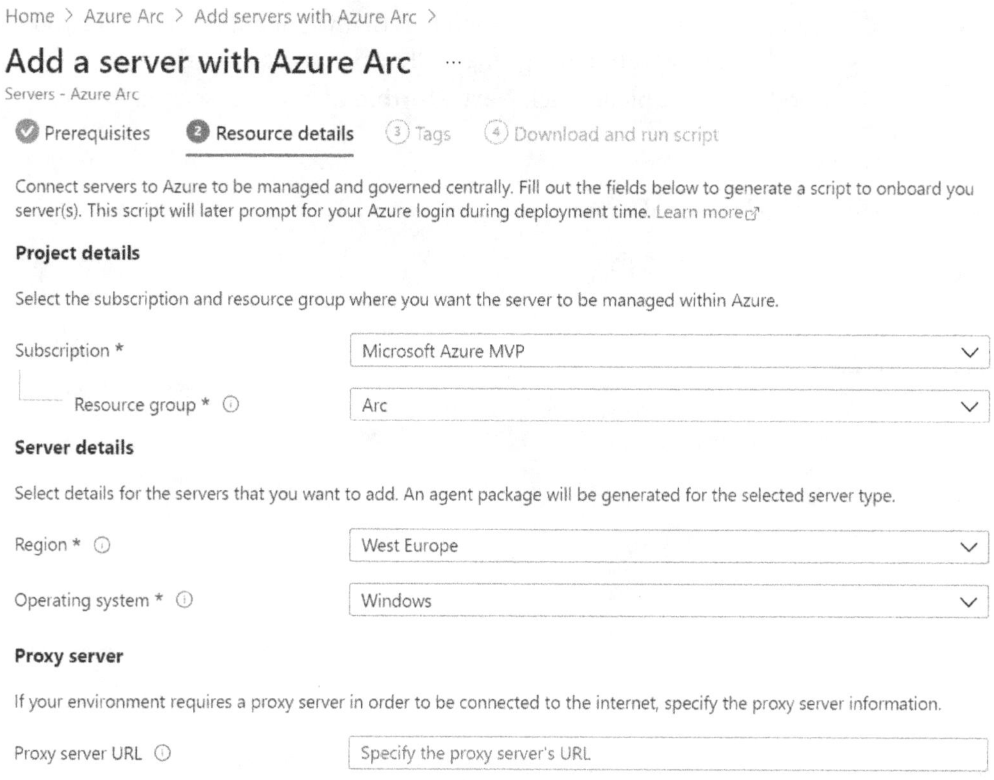

Home > Azure Arc > Add servers with Azure Arc >

Add a server with Azure Arc ⋯
Servers - Azure Arc

✅ Prerequisites   ② **Resource details**   ③ Tags   ④ Download and run script

Connect servers to Azure to be managed and governed centrally. Fill out the fields below to generate a script to onboard you server(s). This script will later prompt for your Azure login during deployment time. Learn more↗

Project details

Select the subscription and resource group where you want the server to be managed within Azure.

Subscription * | Microsoft Azure MVP   ⌄ |

Resource group * ⓘ | Arc   ⌄ |

Server details

Select details for the servers that you want to add. An agent package will be generated for the selected server type.

Region * ⓘ | West Europe   ⌄ |

Operating system * ⓘ | Windows   ⌄ |

Proxy server

If your environment requires a proxy server in order to be connected to the internet, specify the proxy server information.

Proxy server URL ⓘ | Specify the proxy server's URL |

Figure 2.3 – Add a server with Azure Arc – Resource details

7. On the **3** **Tags** blade, enter physical location tags that will identify the server you are adding, as well as any additional tags that will help you organize the resources better. At the bottom of the blade, click **Next: Download and run script**:

Home > Azure Arc > Add servers with Azure Arc >

Add a server with Azure Arc   ...
Servers - Azure Arc

✓ Prerequisites   ✓ Resource details   **3** **Tags**   ④ Download and run script

To manage and create custom views of your resources, assign tags. Learn more about tags↗

Physical location tags

Start with these options for physical location types, change them to suit your needs, or create your own. If you leave the value field blank for these options, the tags will not be created.

Name ⓘ	Value ⓘ	Resource	
Datacenter	: Sta-5-Kis	Server - Azure Arc	🗑
City	: Stockholm	Server - Azure Arc	🗑
Region	: Nordics	Server - Azure Arc	🗑
Country	: Sweden	Server - Azure Arc	🗑
	:	Server - Azure Arc	

Custom tags

Add additional tags to help you organize your resources to facilitate administrative tasks.

Name ⓘ	Value ⓘ	Resource	
Department	: Development	Server - Azure Arc	🗑
Type	: Testing	Server - Azure Arc	🗑
	:	Server - Azure Arc	

Figure 2.4 – Add a server with Azure Arc – Physical location tags

8. In the ❹ **Download and run script** section, click on the **Download** button to download and save the `OnboardingScript.ps1` script. Copy the script onto the server you are onboarding to Azure Arc. Select **Close** to finish and close the **Add a server with Azure** blade. On the top-right side of the **Add servers with Azure Arc** blade, click **X** to close the blade and return to the **Azure Arc | Servers** blade:

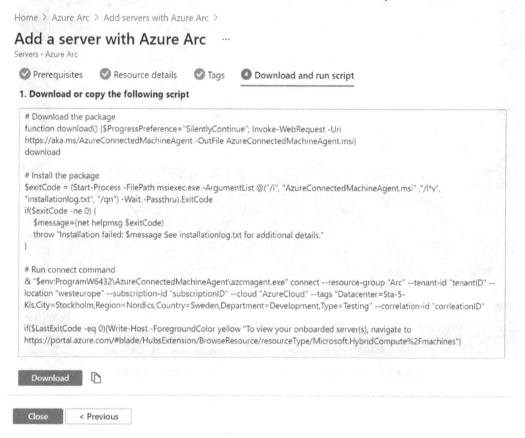

Figure 2.5 – Add a server with Arc – Download and run script

9. On the server you are onboarding to Azure Arc, run `OnboardingScript.ps1`. The script will download and install **Azure Connected Machine Agent**, initiate authentication to Azure, create an Azure Arc-enabled server resource, and associate it with the agent:

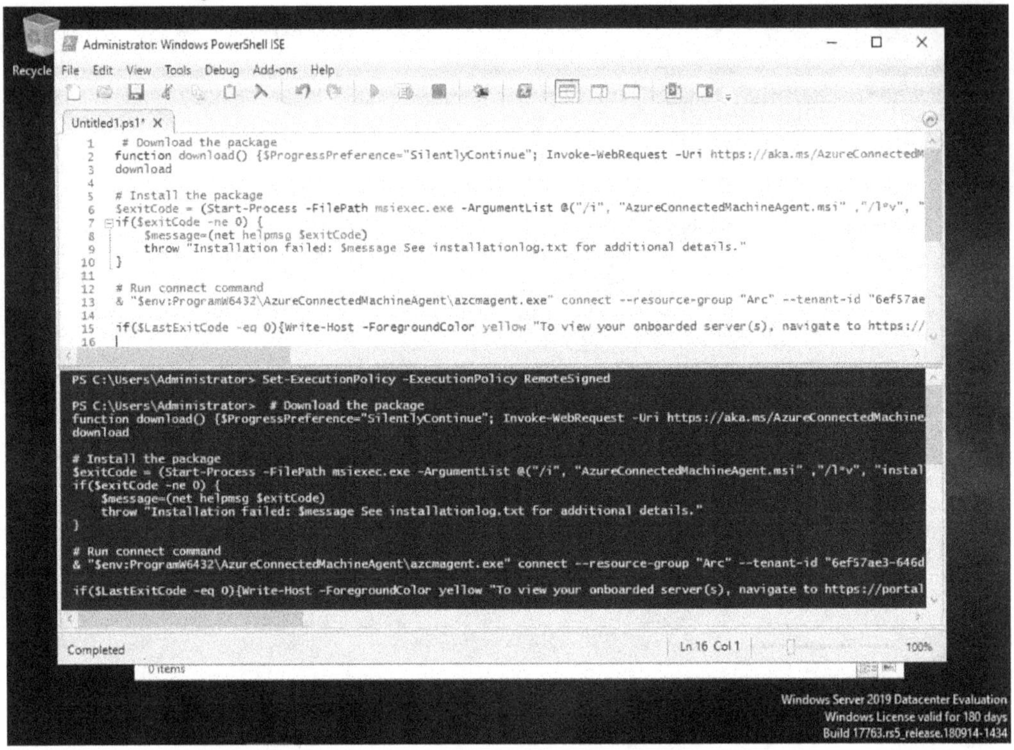

Figure 2.6 – Executing the installation script on a target machine

10. After successfully installing **Azure Connected Machine Agent**, authenticating, and creating an Azure Arc enabled resource in Azure, you should receive a message about having signed into the **Azure Connected Machine Agent** application:

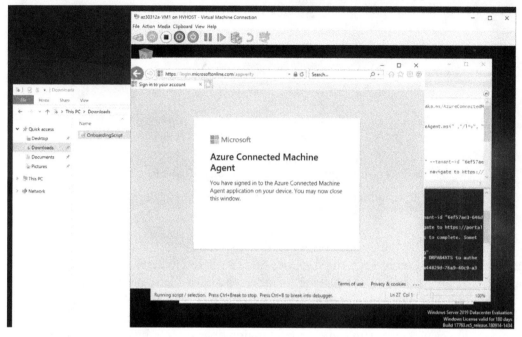

Figure 2.7 – Successful Azure Connected Machine Agent sign-in message

11. In the Azure portal, on the **Azure Arc | Servers** blade, click **Refresh** to display the newly onboarded server in Azure Arc:

Figure 2.8 – Newly onboarded server visible in the Azure Arc | Servers blade

12. After a while, the newly added Azure Arc server will be visible in Microsoft Defender for Cloud, including its **Inventory** and **Recommendations**:

Figure 2.9 – Newly onboarded server visible in Microsoft Defender for Cloud | Inventory

13. Click on the newly added server. The **Resource Health** page opens.

How it works...

Azure Arc-enabled servers is a cloud service that allows you to manage servers hosted outside of Azure, such as on-premises or on other cloud providers, and it is the preferred way of adding non-Azure machines to Microsoft Defender for Cloud. Azure Arc-enabled machines support additional monitoring and configuration management tasks, such as configuration changes reporting, guest configuration policies, VM Insights, simplified deployment, update management, security monitoring, threat detection, and others. Once you connect a non-Azure machine to Azure Arc, it will be visible and protected by Microsoft Defender for Cloud.

Connecting non-Azure virtual machines using Microsoft Defender for Cloud portal pages

Before Microsoft Defender for Cloud can monitor your security posture and display security recommendations for your non-Azure computers, you must connect them to Azure. This recipe will show you how to connect a non-Azure server using Microsoft Defender for Cloud portal pages.

Getting ready

Before you start connecting servers to Azure using Microsoft Defender for Cloud portal pages, you must have administrative permissions on a target server to install and configure it.

Open a web browser and navigate to `https://portal.azure.com`.

How to do it...

To enable Microsoft Defender for Cloud on multiple subscriptions at once, complete the following steps:

1. In the Azure portal, open **Microsoft Defender for Cloud**. You can open Microsoft Defender for Cloud in multiple ways: typing **Microsoft Defender for Cloud** in a search bar, clicking on a link to it via **Favorite**, or by going to **All Services**. On the left-hand side menu, select **Inventory**:

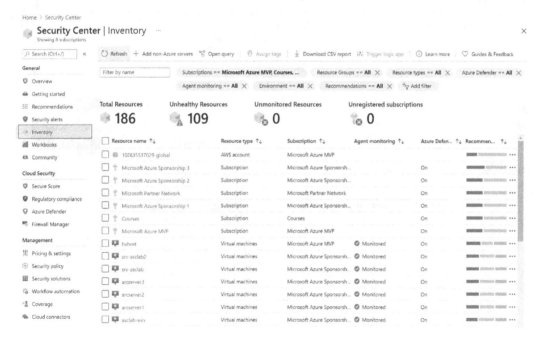

Figure 2.10 – Defender for Cloud – Inventory

2. On the **Onboard servers to Security Center** page, you will see the list of Log Analytics workspaces that you have permission for. Click on **Upgrade** next to the workspace name where you want to store the data. Otherwise, if you want to create a new workspace or if there are no Log Analytics workspaces available, click on **Create New Workspace**:

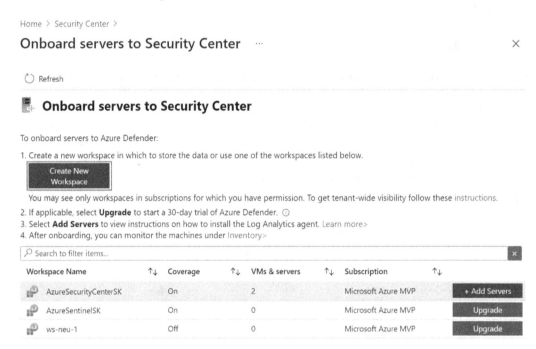

Figure 2.11 – The Onboard servers to Security Center page – Add Servers

3. On the right-hand side of the **Onboard server to Security Center** page, click on the **+ Add Servers** button:

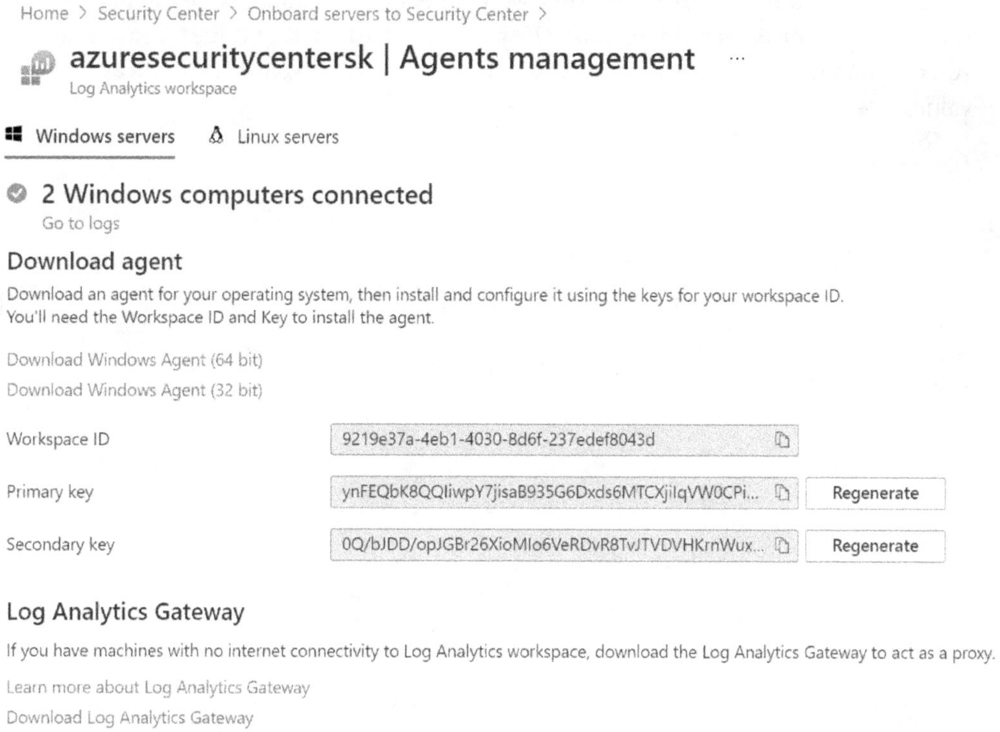

Home > Security Center > Onboard servers to Security Center >

azuresecuritycentersk | Agents management   ...
Log Analytics workspace

■ Windows servers   ⚊ Linux servers

✓ **2 Windows computers connected**
Go to logs

Download agent

Download an agent for your operating system, then install and configure it using the keys for your workspace ID.
You'll need the Workspace ID and Key to install the agent.

Download Windows Agent (64 bit)
Download Windows Agent (32 bit)

Workspace ID	9219e37a-4eb1-4030-8d6f-237edef8043d	
Primary key	ynFEQbK8QQIiwpY7jisaB935G6Dxds6MTCXjilqVW0CPi...	Regenerate
Secondary key	0Q/bJDD/opJGBr26XioMlo6VeRDvR8TvJTVDVHKrnWux...	Regenerate

Log Analytics Gateway

If you have machines with no internet connectivity to Log Analytics workspace, download the Log Analytics Gateway to act as a proxy.

Learn more about Log Analytics Gateway
Download Log Analytics Gateway

Figure 2.12 – Log Analytics workspace connection – Agents management

4. On the workspace **Agents management** page, download and install an appropriate version (64-bit or 32-bit) of **Windows Agent (Microsoft Monitoring Agent)** on a machine you are onboarding to Microsoft Defender for Cloud. Note the fields showing the **Workspace ID**, **Primary Key**, and **Secondary Key** values. Leave the **Agents Management** page open as you will need the **Workspace ID** and **Key** values later:

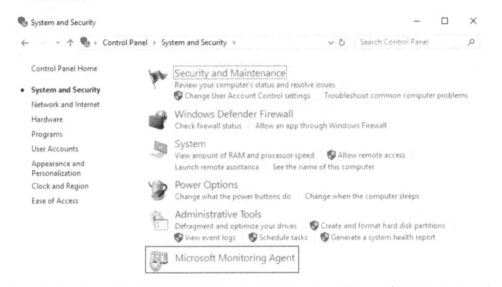

Figure 2.13 – Control Panel displaying the installed Windows Agent (Microsoft Monitoring Agent)

5. After successfully installing **Windows Agent (Microsoft Monitoring Agent)**, open **Control Panel** and click **System and Security**. Click on **Microsoft Monitoring Agent** to run the application:

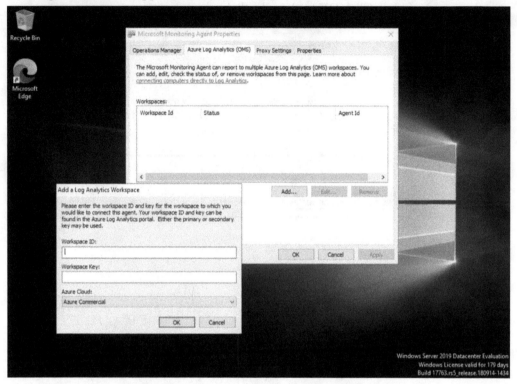

Figure 2.14 – Microsoft Monitoring Agent Properties

6. On the **Microsoft Monitoring Agent Properties** window, select the **Azure Log Analytics (OMS)** tab and click the **Add…** button. The **Add a Log Analytics Workspace** dialog box will open.

7. At the **Add a Log Analytics Workspace** dialog box, enter the **Workspace ID** value and either the primary or secondary **Workspace Key**. Both *Step 4* and *Figure 2.12* refer to the **Agent Management** page in the Azure portal, which is where these values can be found. Click **OK** to close the dialog box:

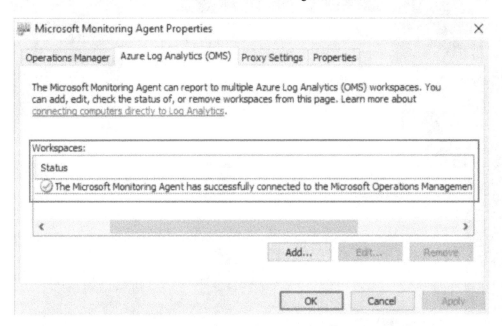

Figure 2.15 – Microsoft Monitoring Agent Properties window showing a successfully connected message

8. After a few moments, on the **Azure Log Analytics (OMS)** tab, an entry in the **Workspaces** field should show a message confirming that a successful connection has been established to a Log Analytics workspace in Azure. Click **OK** to close this window.

9. In a browser, in the Azure portal, in the search bar, type **Log Analytics** and click on a **Log Analytics workspaces** entry in the search results. In the newly opened **Log Analytics workspaces** blade, click on the workspace name you selected in *Step 3*; that is, the workspace you chose to store the data in.

10. From the left menu, in the **General** section, select **Logs**:

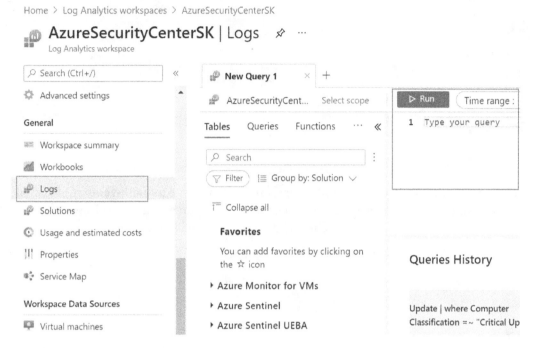

Figure 2.16 – Log Analytics workspace – The Logs blade

11. On the right, type and run the following **Kusto Query Language** (**KQL**) query to check if a machine you onboarded is connected to the workspace:

```
Heartbeat
| where OSType == 'Windows'
| summarize arg_max(TimeGenerated, *) by SourceComputerId
| sort by Computer
| render table
```

12. The query results list should contain the name of the newly onboarded (that is, connected) machine:

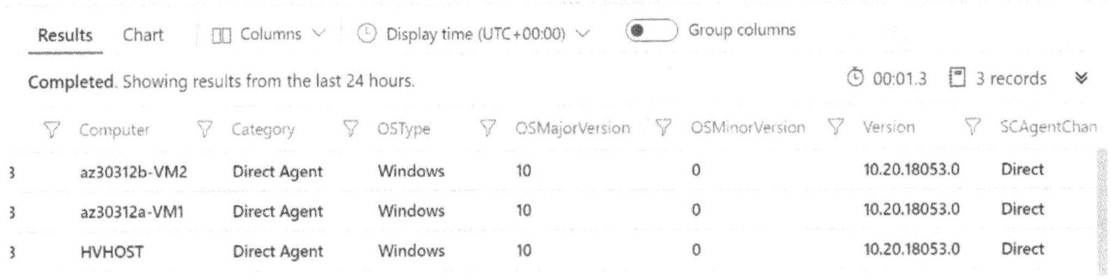

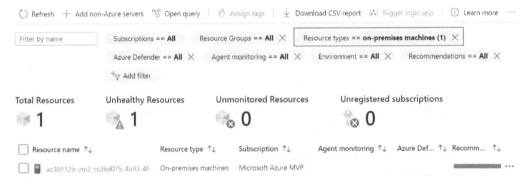

Figure 2.17 – KQL query results showing the newly onboarded machine

13. Finally, open **Microsoft Defender for Cloud** and navigate to the **Inventory** blade. Use an existing filter entry or add a filter-to-filter resource by going to **Resource types**. A newly onboarded machine should be displayed in the inventory list.

Check if Microsoft Defender for Cloud is connected:

Figure 2.18 – Microsoft Defender for Cloud – Inventory list, filtered by Resource types

14. Click on the link representing an on-premises virtual machine. The **Resource Health** page opens.

15. The **Resource Health** page shows resource information and recommendations applicable to the resource, that is, the on-premises virtual machine.

How it works...

Microsoft Defender for Cloud supports adding a non-Azure machine from Security Center's pages in the Azure portal. To complete this, you must enable **Microsoft Defender for Cloud Plans** on a Log Analytics workspace, install Microsoft Monitoring Agent on a target server, and connect it securely to the workspace. To check whether the newly onboarded machine has successfully connected to a Log Analytics workspace, you can use a KQL query and check the presence of a machine in Microsoft Defender for Cloud, on the **Inventory** page.

Setting up Amazon Web Services Config and Amazon Web Services Security Hub

In multi-cloud environments, cloud security services must span multiple cloud platforms as well.

Connecting **Amazon Web Services (AWS)** to **Microsoft Defender for Cloud** requires performing multiple steps. Due to this, we will break the whole process into separate recipes, which will make this easier to understand and implement.

To onboard an AWS account in Microsoft Defender for Cloud, you need to enable **AWS Config** and **AWS Security Hub** first.

Getting ready

Open a browser and navigate to `https://console.aws.amazon.com/`. This recipe presumes you have not already enabled **AWS Config** and **AWS Security Hub**.

How to do it...

To set up **Amazon Web Services Config** and **Amazon Web Services Security Hub**, complete the following steps:

1. In the AWS Console, open **AWS Config**. You can open **AWS Config** in two ways. First, you can type **AWS Config** in a search bar and select that option. Second, you can click on **Services** in the top-left corner and, under **Management & Governance**, select **Config**.

2. If you have never run **AWS Config** before, or if you are configuring AWS Config in a new region, you can choose to click on **Get started** or **1-click setup**. **Get started** will allow you to go through the configuration steps by yourself, while **1-click setup** will auto-complete the setup process based on AWS best practices. In this recipe, we will complete the manual process. Click on **Get started**:

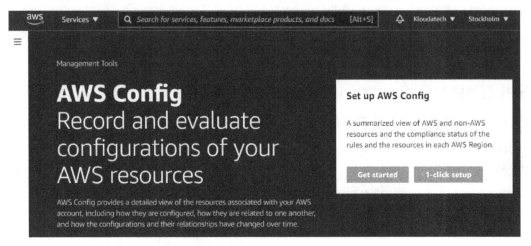

Figure 2.19 – Setting up AWS Config for the first time

3. On the **Step 1 – Settings** page, under the **General settings** section, for **Resource types to record**, select **Record all resources supported in this region**, while for **AWS Config role**, choose **Create AWS Config service-linked role**.

4. In the **Deliver method** section, for **Amazon S3 bucket**, choose **Create a bucket**. It is strongly recommended that you create a unique S3 bucket name as you cannot change the bucket's name once it has been created. Click **Next**:

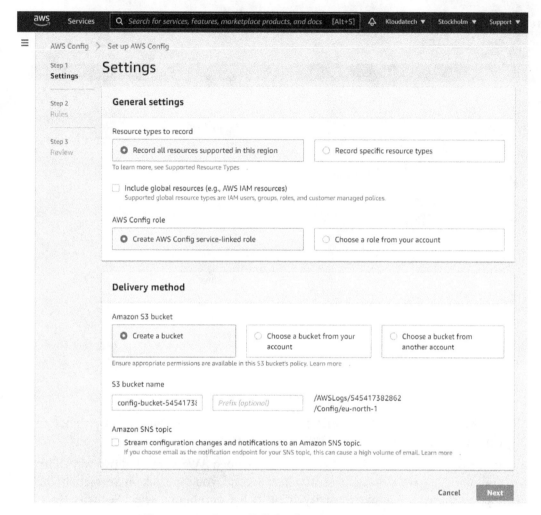

Figure 2.20 – Set up AWS Config – Step 1 – Settings

5.   On the **Step 2 – Rules (AWS Managed Rules)** page, you can add additional AWS managed rules to your account to evaluate your AWS resources against the rules you have chosen. This step is optional. Click **Next**.

6. On the **Step 3 – Review** page, review your AWS Config setup details and
    click **Confirm**:

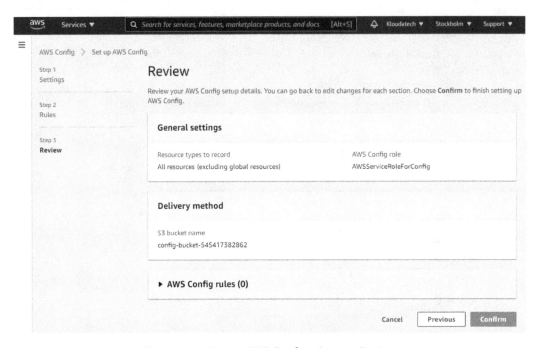

Figure 2.21 – Set up AWS Config – Step 3 – Review

7. On the **Welcome to AWS Config** screen, click on the **X** button at the top right to
    close the window and display **AWS Config console**.

8. In **AWS Console**, open **AWS Config**. You can open **AWS Config** in two ways. First,
    you can type **AWS Config** in a search bar. Second, you can click on **Services** in the
    top-left corner and, under **Security, Identity & Compliance**, select **Security Hub**.

9. **AWS Security Hub – Get started with Security Hub** will appear if you have never
    run **AWS Security Hub** before. If you have not run **AWS Security Hub** before, click
    on **Go to Security Hub**. Otherwise, skip to *Step 10*.

10. The **Welcome to AWS Security Hub** page will open. Examine the **Security
    Standards** section. Select the checkboxes next to all the security standards and click
    Enable Security Hub. Once you've enabled Security Hub, it can take up to 2 hours
    to see the results from security checks in AWS Security Hub.

How it works...

Amazon Web Services Config enables you to view the configuration of your AWS resources in detail, track the configuration of resources, retrieve historical configuration data, receive notifications about resource life cycle-related events, view relationships between resources, and more. **Amazon Web Services Security Hub** is a security center for AWS resources. It collects security-related data from AWS resources from supported AWS Partner Network security solutions, as well as supported AWS services such as Amazon Macie, Amazon GuardDuty, and Amazon Inspector. AWS Security Hub and Microsoft Defender for Cloud will use AWS Config data for the inventory and security statuses of AWS resources.

Creating an Identity and Access Management AWS role for Microsoft Defender for Cloud

To enable Microsoft Defender for Cloud to connect to and allow it to authenticate to AWS, you have two options: create an AWS user for Microsoft Defender for Cloud or create an AWS **Identity and Access Management (IAM)** role for Microsoft Defender for Cloud. The first option is less secure, while the second option is the most secure and preferred way to authenticate Microsoft Defender for Cloud to AWS. In this recipe, you will use the more secure option to connect an AWS account to Microsoft Defender for Cloud, you will create an IAM role.

Getting ready

Open a browser and navigate to `https://console.aws.amazon.com/`. Open a new tab in a web browser and navigate to `https://portal.azure.com`.

How to do it...

To create an **Identity and Access Management (IAM)** role for Microsoft Defender for Cloud, complete the following steps:

1. In the Azure portal, open **Defender for Cloud**. From the menu, under the **Management** section, select **Environment settings**. Switch back to the classic cloud connectors experience. On the **Cloud connectors** blade, from the top menu, select **Connect AWS account**:

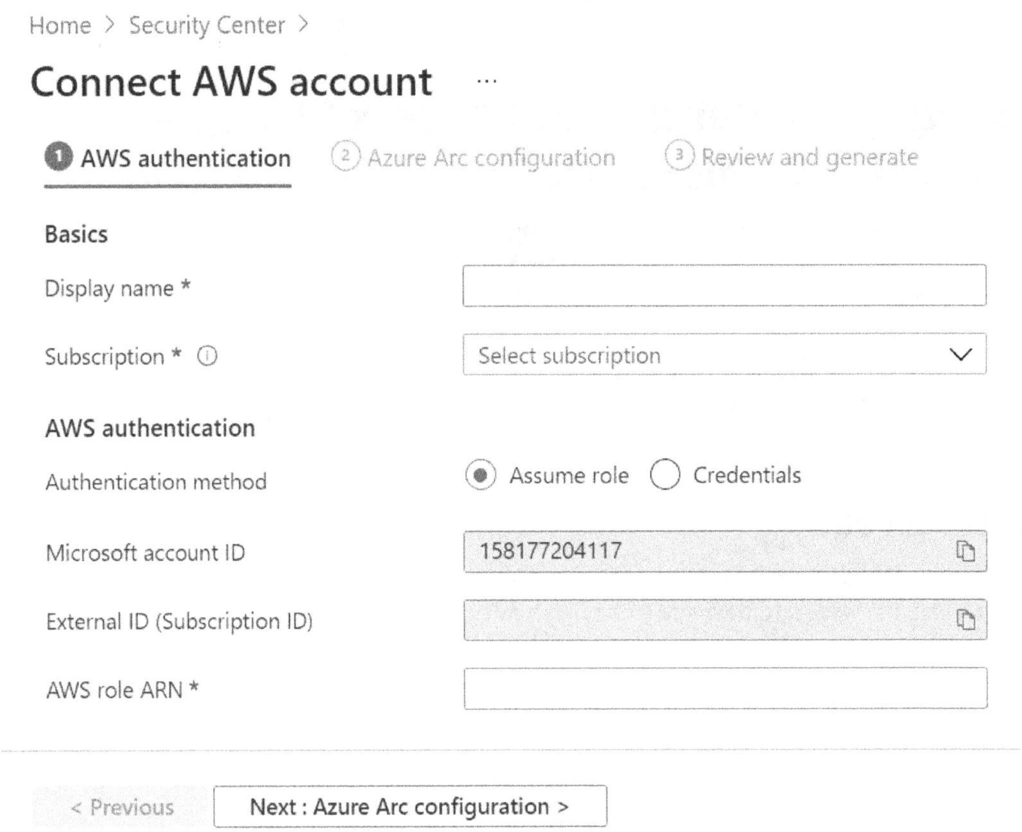

Home > Security Center >

Connect AWS account   ...

1 AWS authentication	② Azure Arc configuration	③ Review and generate

Basics

Display name *

Subscription * ⓘ   Select subscription   ⌄

AWS authentication

Authentication method   ◉ Assume role   ◯ Credentials

Microsoft account ID   158177204117

External ID (Subscription ID)

AWS role ARN *

< Previous   **Next : Azure Arc configuration >**

Figure 2.22 – Connect AWS account

2. In the **Display name** field, enter a name to identify the AWS account connection in Microsoft Defender for Cloud. From the **Subscription** menu, choose an Azure subscription. Take note of the **Microsoft account ID** and **External ID (Subscription ID)** values as you will need them in the next steps. Leave the Azure portal browser tab open.

3. Switch to the **AWS Console** browser tab. In the AWS Console, open **IAM**. You can open **IAM** in two ways. First, you can type **IAM** in a search bar and select it. Second, you can click on **Services** in the top-left corner and, under **Security, Identity, & Compliance**, select **IAM**.

4. On the left-hand side menu, select **Roles** and then **Create Role**:

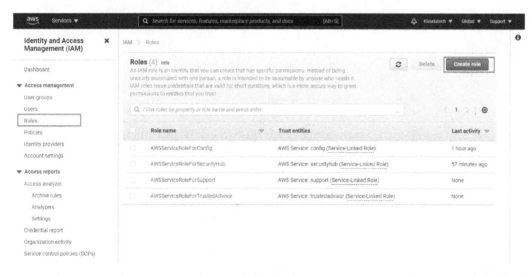

Figure 2.23 - Identity and Access Management (IAM) console

5. On the **Create role** page, select **Another AWS Account**. This step requires the values that we specified in *Step 2*. In the **Account ID** field, enter 158177204117 as your Microsoft Account ID. Select the **Require external ID** checkbox. For **External ID**, enter the **Microsoft Azure subscription ID** value. Click **Next: Permissions**:

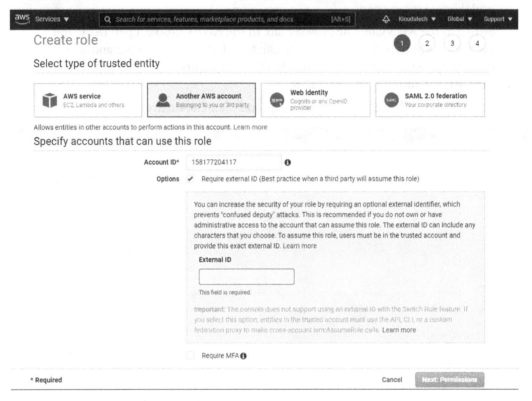

Figure 2.24 – Create role – Specify accounts that can use this role

6. On the **Attach permission policies** page, select the **SecurityAudit**, **AmazonSSMAutomationRole**, and **AWSSecurityHubReadOnlyAccess** policies. Then, click **Next: Tags**.

7. Optionally, you can add **Tags**. Click **Next: Review**.

8. On the **Review** page, in the **Role Name** field, enter the name of the role you just created. Optionally, in the **Role Description** field, enter some text that will describe the newly created role. Click **Create role** to finish creating the role.

9. On the **Identity and Access Management (IAM)** page, click on the role name you created.

10. On the **Summary** page, copy the **Role ARN** value, as you will need it in the following recipe.

How it works...

To be able to connect your AWS account to **Microsoft Defender for Cloud**, Microsoft Defender for Cloud needs to have permission to access information about AWS resources. For this, you have created an AWS IAM role and attached the appropriate permission policies to it. The policies grant various permissions, including read-only access to AWS Security Hub, and read access to information about AWS services and AWS Systems Manager Agent automation permissions. Later, you will reference this role while connecting your AWS account to Microsoft Defender for Cloud.

Connecting Amazon Web Services to Microsoft Defender for Cloud

The final step for connecting an **AWS** account to **Microsoft Defender for Cloud** is to create an AWS connector in Microsoft Defender for Cloud.

Getting ready

Open a web browser and navigate to `https://portal.azure.com`. Open a new browser tab and navigate to `https://console.aws.amazon.com/`.

How to do it...

To connect AWS to Microsoft Defender for Cloud and create an AWS connector in Microsoft Defender for Cloud , complete the following steps:

1. Open **Microsoft Defender for Cloud**. From the left menu, under the **Management** section, select **Environment settings**. Switch back to the classic cloud connectors experience. On the **Cloud connectors** blade, from the top menu, select **Connect AWS account**. In the **Display name** field, enter a name to identify the AWS account connection in Microsoft Defender for Cloud. From the **Subscription** menu, choose an Azure subscription.

2. In the **AWS role ARN** field, paste the **Role ARN** value you copied in the *Creating an Identity and Access Management (IAM) Amazon Web Services (AWS) role for Microsoft Defender for Cloud* recipe. Click **Next: Azure Arc configuration**.

3. Alternatively, if you do not have a **Role ARN** string ready, in **AWS Console**, in the top-left corner, click **Services** and, under the **Security, Identity, and Compliance** section, select **IAM**.

4. From the left menu, click **Roles** and click on the name of an AWS role you created to connect to Microsoft Defender for Cloud. Copy the **Role ARN** value.

5. Switch back to the **Azure portal** browser tab. In the **AWS role ARN** field, paste the **Role ARN** value. Click **Next: Azure Arc configuration**.

6. On the **Connect AWS account – ❷ Azure Arc configuration** page, choose or create a new **Resource group**, and then choose a **Region**.

7. In the **Authentication** section, click on the **Create a Service Principal in Azure Active directory with Azure Connected Machine Onboarding role with a few clicks** link:

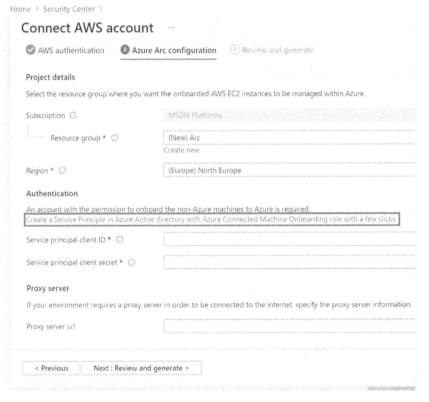

Figure 2.25 – Connect AWS account – ❷ The Azure Arc configuration page

8. Copy the script and open **Cloud Shell**. Paste and execute the script in Azure PowerShell. The script will return a GUID, or an **Application (client) ID** value, and a **Password** value.

9. Paste the **Application (client) ID** value into the **Service principal client ID** field. Paste the **Password** value into the **Service principal client secret** field. Then, click **Next: Review and generate**.

10. On the **Connect AWS account – ❸ Review and generate** page, review the configuration details and click **Create**.

11. In **Microsoft Defender for Cloud**, on the **Cloud Connectors** page, you should see the newly added AWS connector. Its **Status** should be **Valid** if it connected successfully:

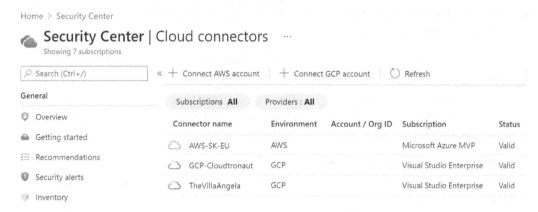

Figure 2.26 – Microsoft Defender for Cloud – Cloud Connectors

12. After a few minutes, in **Microsoft Defender for Cloud**, the **Regulatory Compliance** section will display the AWS compliance controls, while the AWS resources and recommendations will be visible in the **Inventory** and **Recommendations** sections:

Cloud Security	Recommendation	Unhealthy resources ↑↓	Resource health ↑↓	Initiative ↑↓
Secure Score				
Regulatory compliance	Avoid the use of the "root" account	1 of 1 AWS resources		AWS CIS 1.2.0
Azure Defender	S3 Block Public Access setting should be enabled	1 of 1 AWS resources		AWS-Foundational-Security...
Firewall Manager	Password policies for IAM users should have strong configurations	1 of 1 AWS resources		AWS-Foundational-Security...
Management	Hardware MFA should be enabled for the root user	1 of 1 AWS resources		AWS-Foundational-Security...
Pricing & settings	GuardDuty should be enabled	1 of 1 AWS resources		AWS-Foundational-Security...
Security policy	Amazon EC2 should be configured to use VPC endpoints	4 of 4 AWS resources		AWS-Foundational-Security...
Security solutions	EBS default encryption should be enabled	1 of 1 AWS resources		AWS-Foundational-Security...
Workflow automation	The VPC default security group should not allow inbound and outbound t...	4 of 4 AWS resources		AWS-Foundational-Security...
Coverage	AWS Config should be enabled	1 of 1 AWS resources		AWS-Foundational-Security...
Cloud connectors	CloudTrail should be enabled and configured with at least one multi-Regi...	1 of 1 AWS resources		AWS-Foundational-Security...
	CloudTrail should be enabled	1 of 1 AWS resources		AWS-PCI-DSS-3.2.1

Figure 2.27 – Microsoft Defender for Cloud – AWS Recommendations

How it works...

The final step in connecting an AWS account to Microsoft Defender for Cloud is to create an AWS connector in Microsoft Defender for Cloud. You need to associate an AWS role with an Azure subscription and create a service principal that will be used to authenticate access to Azure. Once you connect your AWS account to Microsoft Defender for Cloud, you can use Security Center's capabilities to protect AWS account assets in Microsoft Defender for Cloud.

There's more...

If you close Cloud Shell and do not know the **Service principal client ID** and **Service principal client secret** values anymore, or you want to change the current secret value of the application, complete the following steps:

1. In the Azure portal, open **Azure Active Directory**. From the left menu, under the **Manage** section, select **App registrations**.

2. On the **App registrations** blade, click the **All applications** tab. In the **Search** field, type Arc to filter the application entries. In the **Application (client) ID** column, identify the string to use as a **Service principal client ID** value.

3. Click on the application name, then select **Certificates & secrets**.

4. Under **Client secrets**, click on the recycle bin icon to delete a **Secret** and click + **New client secret**.

5. Enter a **Description** and choose an expiration time. Then, click **Add**.

6. Under **Client secrets**, copy the string representing the **Value** property of the secret. This will be used as a **Service principal client secret** value.

Configuring GCP Security Command Center and enabling GCP Security Command Center API

For any environment that spans multiple cloud providers, cloud security services must span multiple cloud platforms as well.

Connecting a **Google Cloud Platform** (**GCP**) environment to **Microsoft Defender for Cloud** involves several steps. We will break this process into separate recipes so that this will be easier to understand and implement.

To onboard a **Google Cloud Platform** account into **Microsoft Defender for Cloud**, you need to configure **GCP Security Center** and enable **Security Health Analytics** first.

Getting ready

Before configuring **GCP Security Center**, you should have **GCP Organization** and a **Google Cloud Identity** account set up.

Open a web browser and navigate to https://console.cloud.google.com.

How to do it...

To onboard a GCP account into Microsoft Defender for Cloud, complete the following steps:

1. In **GCP Console**, in the top-right corner select, an account. The selected account should belong to a GCP organization that contains or will contain a project you will connect to **Security Command Center**. If you already have a project, skip to *Step 4*.

2. If the dashboard area is empty and you do not have a project, click **Create project**.

3. On the **New Project** page, enter a value for **Project name**. Choose the project's **Organization** and **Location** and click **Create**.

4. On the left-hand side menu, under **Security**, click **Security Command Center**. If you get a message stating **Page not viewable for projects**. This page is only viewable in the project/folder scope for Premium Tier organizations. Upgrade your organization to Premium, then from the drop-down menu on the right, select an **Organization** and click **Select**.

5. If you get an error message stating **You do not have sufficient permissions to view this page**, from the top-left corner, click **Google Cloud Platform**. Then, under **IAM & Admin** menu, select **IAM**:

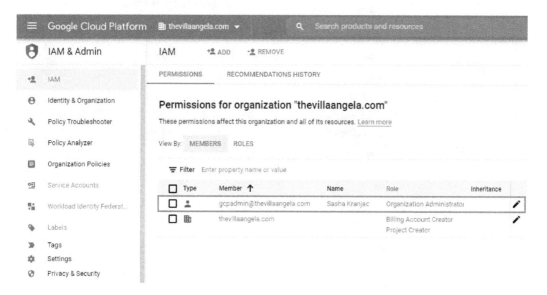

Figure 2.28 – Editing the service account role permissions

6. Identify the account with an **Organization Administrator** role or the account that you are currently logged into and using to set up **Security Command Center**. To edit the account, click the pencil icon next to it.

7. Click + **ADD ANOTHER ROLE** to add a role. Add the **Security Center Admin**, **Security Admin**, and **Create Service Account** roles and click **SAVE**:

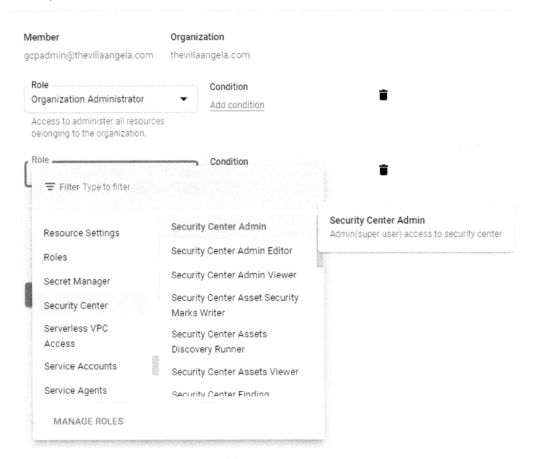

Figure 2.29 – Adding roles to a service account

8. If you receive a message stating that Security Command Center has not been onboarded or activated yet, refresh the browser tab. Make sure you selected the right organization. After few moments, the **Settings – ❶ Get started** page should open, and **Security Command Center**, **Standard** tier should be selected. Click **NEXT**:

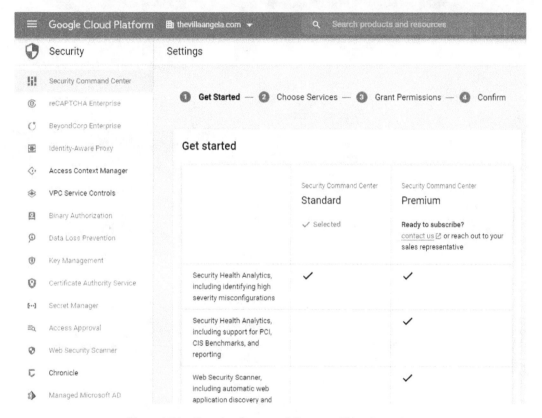

Figure 2.30 – Security Command Center – Choosing a tier

9. On the **❷ Choose Services** page, check whether **Security Health Analytics** is **Enabled** by default, and review the rest of the information. Click **NEXT**.

10. On the ❸ **Grant Permissions** page, review the **Required Roles** and **Service Account Created** information. Click **GRANT ROLES**. The messages should indicate if the process of granting roles and the test have been completed. Click **NEXT**:

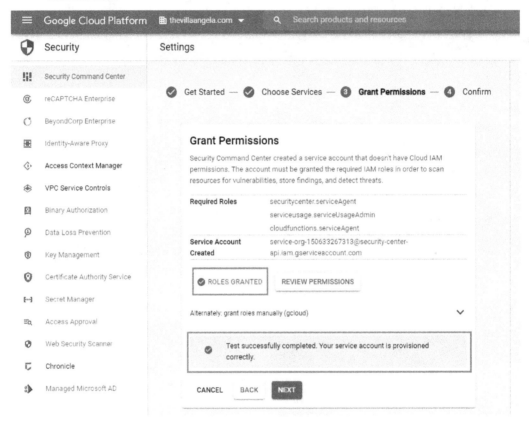

Figure 2.31 – Granting permissions to a service account

11. On the ❹ **Confirm** page, **Ready to complete setup** will inform you that you are ready to finish setting up **Security Command Center**. Click **FINISH** to complete the **Security Command Center** setup process.

12. Click the **Vulnerabilities** tab to display **Security Health Analytics** findings for the organization. To display **Security Health Analytics** for a project, under **Projects Filter**, click the plus (+) sign to **Add a project to the Projects Filter**.

13. In **GCP Console**, under **APIs & Services**, click **Dashboard**. Click on **Library** or + **ENABLE APIS AND SERVICES**.

14. On the **Welcome to the API Library** page, in the **Search for APIs & Services** search bar, type `security`.

15. Click on an entry representing **Security Command Center API**:

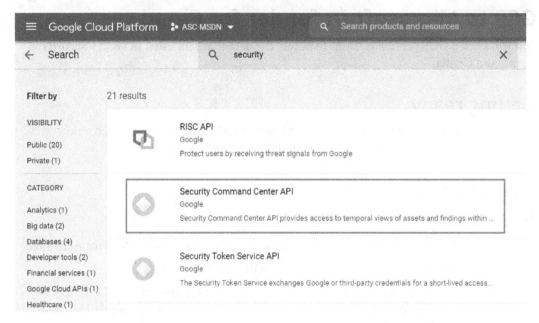

Figure 2.32 – Choosing a Security Command Center API

16. On the **Security Command Center API** page, click **ENABLE**.

How it works...

The first part of connecting a GCP account to Microsoft Defender for Cloud is to set up Security Command Center for your organization. First, you must grant appropriate permissions to an administrative account. This way, Security Command Center will display security-related information about services. Once you've completed the necessary setup, it may take some time for security information to be displayed in Security Command Center. To enable Security Command Center's functionality, you must provide access to cloud assets and findings within an organization via the **Security Command Center API**.

Creating a GCP service account and connecting GCP to Microsoft Defender for Cloud

You will need a GCP service account to access the data that's ingested in GCP's Security Command Center.

Getting ready

Open a web browser and navigate to `https://portal.azure.com`. Open a new browser tab and navigate to `https://console.cloud.google.com`.

How to do it...

To create a GCP service account, complete the following steps:

1. In **GCP Console**, from the left menu, under **IAM & Admin**, select **Service Accounts**.

2. From the top menu, select **+ CREATE SERVICE ACCOUNT**.

3. In the **Service account name** field, enter the account's name. Optionally, in the **Service account description** field, describe what this service account will be used for. Click **CREATE AND CONTINUE**.

4. When you get to the ❷ **Grant this service account access to project** step, click on a **Select role** drop-down menu and select the **Security Center Admin Viewer** role. Select **CONTINUE**.

5. The **Grant users access to this service account** step is optional. Click **DONE**.

6. Copy the email of the service account and save it; we will use this later.

7. From the **Navigation** menu, under **IAM & Admin**, click **IAM**. Ensure you are viewing IAM permissions for an organization, rather than for a project. From the top menu, switch to the organization level.

8. To add a new user, click **ADD**.

9. In the **New members** field, enter the **email** value of the service account you copied in *Step 6*.

10. From the **Select role** menu, select the **Security Center Admin Viewer** role and click **SAVE**.

11. Select a project to switch to a project level since the **Service Accounts** page can't be viewed by organizations.

12. In the **Navigation** menu, under **IAM & Admin**, click **Service accounts**.

13. Next to the service account you created previously, click on the vertical ellipsis (three vertical dots) and select **Manage keys**:

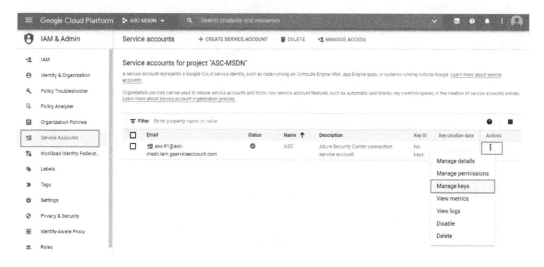

Figure 2.33 – Service accounts for project

14. On the **Keys** page, click **ADD KEY** and then select **Create new key**.

15. In the **Create private key** window, choose **JSON** for **Key type** and click **CREATE**.

16. Save the JSON file.

17. In the **Navigation** menu, under **IAM & Admin**, click **Settings**. Copy the **Organization ID** value and save it for later use.

18. In the Azure portal, open **Microsoft Defender for Cloud**. Under the **Management** section, select **Cloud connectors**.

19. Select **Connect GCP account**:

Figure 2.34 – Connect GCP account

20. In the **Display name** field, enter a name for the GCP connector.

21. Choose an Azure subscription.

22. In the **Organization ID** field, paste the value you copied in *Step 17*.

23. For **GCP private key file**, click on a folder icon and select the JSON file you saved in *Step 16*.

24. Select **Next: Review and generate**.

25. On the ❷ **Review and generate** page, review the details and click **Create**.

How it works...

To connect the GCP account to Microsoft Defender for Cloud, you created a service account in GCP with the appropriate permissions to read security information from GCP Security Center. This GCP service account is used to read the data in GCP Security Command Center. Then, you created a private key for the service account, which will be used for authentication in the Microsoft Defender for Cloud GCP connector. With all the necessary information at hand, you created a GCP connector in Microsoft Defender for Cloud.

3
Workflow Automation and Continuous Export

In this chapter, you will learn how to configure **Microsoft Defender for Cloud** workflow automation, configure continuous data export, and automate Microsoft Defender for Cloud responses.

Although automation is great and very useful, since it reduces security administrator fatigue and greatly shortens the time to react to events and potential threats, it is of the utmost importance to know that you should not automate everything immediately. Automation needs to be planned, and a decision to automate, especially if it concerns security, needs to be justified and developed over time.

Enabling automation just to reduce or eliminate work could make an environment less secure, so some tasks need to remain manual to ensure you can detect potential exposure.

We will cover the following recipes in this chapter:

- Creating logic apps for use in Microsoft Defender for Cloud
- Automating threat detection alert responses
- Automating Microsoft Defender for Cloud recommendation responses
- Automating regulatory compliance standards responses
- Configuring continuous export to Event Hub
- Configuring continuous export to a Log Analytics workspace

Technical requirements

To complete the recipes in this chapter, you will need the following:

- An Azure subscription
- An Office 365 email account
- A web browser, preferably Microsoft Edge

The code samples for this chapter can be found at `https://github.com/PacktPublishing/Microsoft-Defender-for-Cloud-Cookbook`.

Creating logic apps for use in Microsoft Defender for Cloud

Azure Logic Apps is an important part of automating various actions in Microsoft Defender for Cloud. Logic Apps has a much broader application than just being a part of Microsoft Defender for Cloud automation scenarios, but then, it is important to know how to create a logic app that can be used with Microsoft Defender for Cloud.

This recipe will introduce you to creating a simple Logic App that can be used in Microsoft Defender for Cloud automation. The Logic App will send an email using Office 365 when an Microsoft Defender for Cloud alert is triggered.

Getting ready

Open a web browser and navigate to `https://portal.azure.com`.

How to do it...

To create a Logic App that will send an email when an Microsoft Defender for Cloud alert is triggered, complete the following steps:

1. In the Azure portal, open **Logic Apps**.
2. From the top-left menu, click **+ Add**.
3. On the **Basics** tab, under **Project Details**, choose **Azure Subscription** and **Resource Group** for a Logic App.
4. Under **Instance Details**, click the **Consumption** button to choose a consumption Logic App model.
5. Type in a Logic App name and choose an Azure region that will host the Logic App.
6. Optionally, if a selected Azure region has a Log Analytics workspace, you can enable Log Analytics by selecting **Yes** and choose a workspace where you will store the Logic App data for further analysis.

Home > Logic apps >

Create Logic App   ...

Basics   Tags   Review + create

Create a logic app, which lets you group workflows as a logical unit for easier management, deployment and sharing of resources. Workflows let you connect your business-critical apps and services with Azure Logic Apps, automating your workflows without writing a single line of code.

Project Details

Select a subscription to manage deployed resources and costs. Use resource groups like folders to organize and manage all your resources.

Subscription * ⓘ	Microsoft Azure MVP ⌄
Resource Group * ⓘ	LogicApps ⌄
	Create new

Instance Details

Type *	⦿ Consumption ◯ Standard
	❶ Looking for the classic consumption create experience? Click here
Logic App name *	Send-EmailRecommendation ✓
Region *	North Europe ⌄
Enable log analytics *	⦿ Yes ◯ No
Log Analytics workspace *	ws-neu-1 ⌄

Figure 3.1 – Create Logic App

7. Click **Next: Tags**.

8. Optionally, add **Tags** to categorize the Logic App. Then, click **Review + Create**.

9. Click **Create** to complete the steps and create a Logic App.

10. When the deployment is complete, click **Go to resource**.

11. Under **Templates**, click the **Blank Logic App** tile.

12. In the **Logic App Designer** blade, in the **Search connectors and triggers** field, type Defender for Cloud.

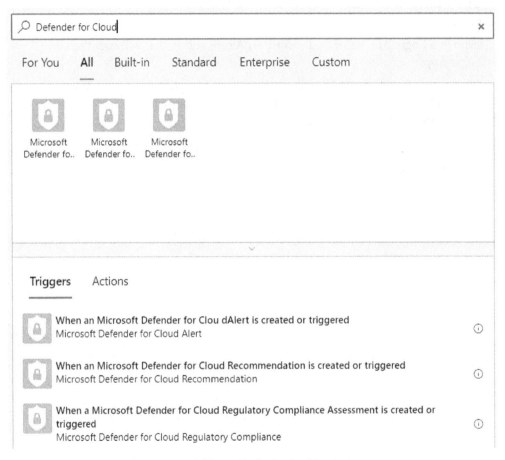

Figure 3.2 – Adding a Defender for Cloud trigger

13. In the lower pane, under the **Trigger** tab, there should be a list of triggers related to Microsoft Defender for Cloud. Click on the **When an Microsoft Defender for Cloud Alert is created or triggered** trigger.

14. In **Logic App Designer**, click **+New step**.

15. In the **Search connectors and triggers** field, type Office 365.

16. From the search results, click on the **Office 365 Outlook** connector.

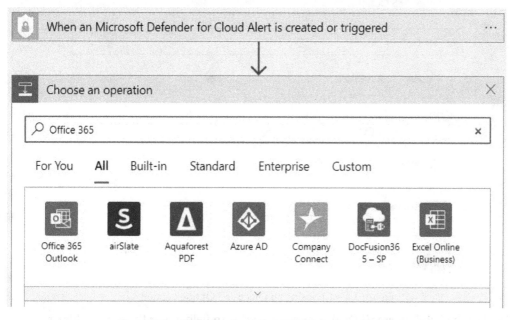

Figure 3.3 – Choosing the Office 365 Outlook connector

17. Under the **Actions** tab, select the **Send an email (V2)** action. You can do this by either scrolling down or typing it in the **Search connectors and actions** field.

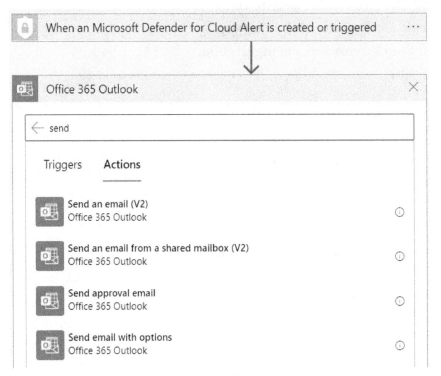

Figure 3.4 – Choosing an Office 365 Outlook action

18. For the **Send an email (V2)** connector, in the **To** field, enter the email addresses separated by semicolons.

19. In the **Subject** field, type Defender for Cloud Alert as the subject of the email.

20. Click in the **Body** field so that the **Add dynamic content** text appears in the bottom-right corner. Click on **Add dynamic content text** once or multiple times until a pop-up window stating **Add dynamic content from the apps and connectors used in this flow** appears on the right.

21. Click **See more** to display the list of actions that are available as dynamic content:

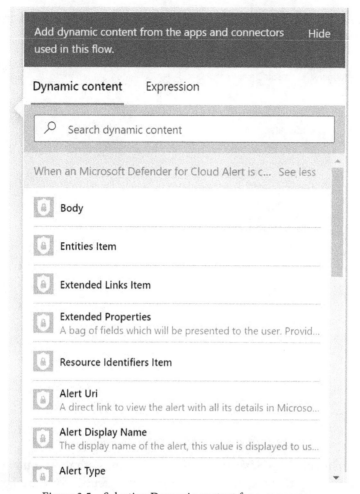

Figure 3.5 – Selecting Dynamic content for a connector

22. Let's construct a sample email body. In the **Body** field, type `Alert Name:`. Then, click the **Alert Display Name** dynamic content field (in the subsequent text, this will be referred to as a *dynamic field*).

23. Type `Alert type:`. Then, select the **Alert Type** dynamic field.

24. Type `Entity:`. Then, select the **Compromised Entity** dynamic field.

25. Type `Description:`. Then, select the **Description** dynamic field.

26. Type `Start time:`. Then, select the **Start Time (UTC)** dynamic field.

27. Type `Severity:`. Then, select the **Severity** dynamic field. The **Send an email (V2)** connector should look like this:

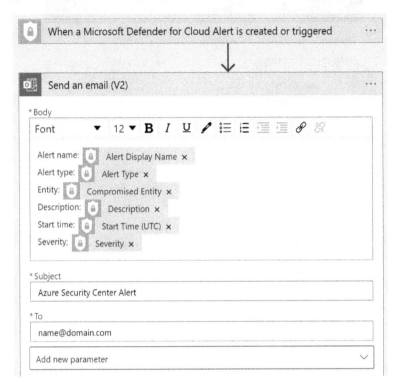

When a Microsoft Defender for Cloud Alert is created or triggered ···

Send an email (V2) ···

* Body

Font   ▼   12 ▼ **B** *I* U̲ ✏ ▤ ▥ ▤ ▤ 𝒫 ⚯

Alert name: 🔒 Alert Display Name ✕
Alert type: 🔒 Alert Type ✕
Entity: 🔒 Compromised Entity ✕
Description: 🔒 Description ✕
Start time: 🔒 Start Time (UTC) ✕
Severity: 🔒 Severity ✕

* Subject

Azure Security Center Alert

* To

name@domain.com

Add new parameter   ⌄

Figure 3.6 – The Send an email (V2) window

28. From the top menu, click **Save** to save the Logic App. With that, you have created a Logic App that can be used in Microsoft Defender for Cloud or other automation scenarios.

How it works...

Logic Apps supports Microsoft Defender for Cloud connectors, which can be used in various automation scenarios. In this simple example, we used an Microsoft Defender for Cloud Logic App connector to provide information about alerts triggered in Microsoft Defender for Cloud and pass alert information to the Office 365 connector to send an email about the triggered alert.

To run the Logic App you have just created, perform the following steps:

1. In the Azure portal, open **Logic Apps**.

2. Click on the **Send-EmailRecommendation** Logic App.

3. On the **Overview** blade, click **Run Trigger** and then **Run**.

4. Under the **Run history** tab, check the status of the Logic App. The **Succeeded** status shows that the Logic App executed successfully.

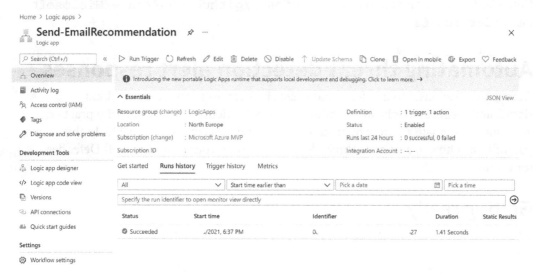

Figure 3.7 – Logic App Overview blade

There's more...

You can click on any Dynamic Content alert to insert a dynamic field into the body of the email. That way, you can insert any information that is relevant to your use case.

At the time of writing this book, Defender for Cloud Connector supports three triggers: **When an Defender for Cloud Alert is created or triggered**, **When an Defender for Cloud Recommendation is created or triggered**, and **When a Defender for Cloud Regulatory Compliance Assessment is created or triggered**.

See also

For the relevant Microsoft Defender for Cloud connector reference, check out the following documentation pages:

- **Security Center Alert**: https://docs.microsoft.com/en-us/connectors/ascalert/

- **Security Center Recommendation**: https://docs.microsoft.com/en-us/connectors/ascassessment/

- **Security Center Regulatory Compliance**: https://docs.microsoft.com/
  en-us/connectors/ascregulatorycomplianceassessment/

More Microsoft Defender for Cloud automation examples are available in the Microsoft Defender for Cloud GitHub repository: https://github.com/Azure/Microsoft-Defender-for-Cloud.

Automating threat detection alert responses

Automating responses to incidents and events in Microsoft Defender for Cloud significantly reduces overhead and administrative burden. As a good security practice, you should automate as many manual responses and procedures as possible. In this recipe, you will learn how to automate threat detection alert responses in Microsoft Defender for Cloud.

Getting ready

Open a web browser and navigate to https://portal.azure.com.

How to do it...

To automate a threat detection alert response in Microsoft Defender for Cloud, complete the following steps:

1. In the Azure portal, open **Microsoft Defender for Cloud**.

2. From the left menu, select **Workflow automation**.

3. From the top menu, click **+ Add workflow automation**. An **Add workflow automation** window will open on the right that has three sections: **General**, **Trigger conditions**, and **Action**.

4. Under the **General** section, in the **Name** field, type in a name without spaces; for example, AntimalwareThreat.

5. In the **Description** field, type in a description of a workflow's automation; for example, Antimalware Threat Response.

6. Select a subscription and a resource group.

7. Under the **Trigger conditions** section, from the **Defender for Cloud data type** menu, select **Security alert**.

8. In the **Alert name contains** field, type the text that will trigger the condition. For this example, type Antimalware.

9. From the **Alert severity** menu, choose **High** and **Medium**.

10. From the **Actions** section, under **Show Logic App instances from the following subscriptions**, choose the subscriptions that you want to display in your Logic Apps.

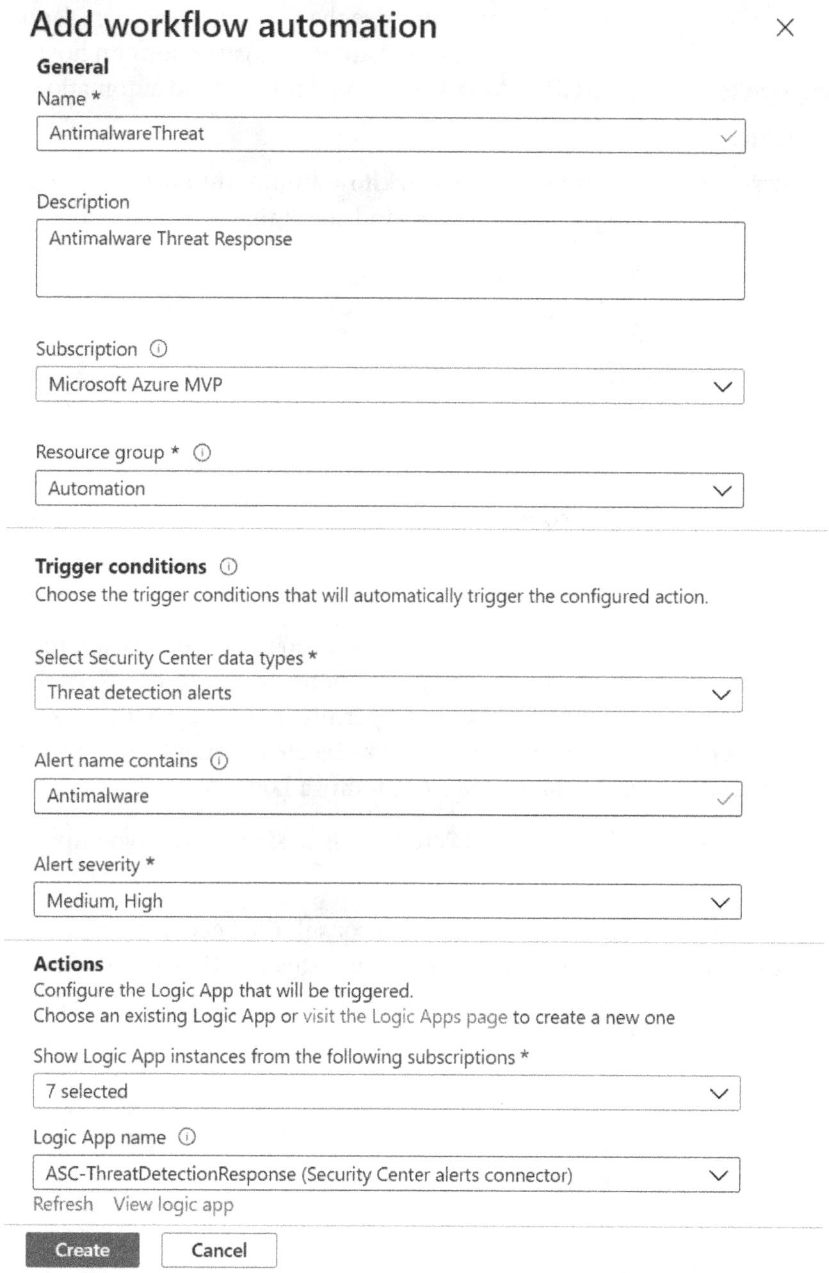

Add workflow automation   ✕

General

Name *

| AntimalwareThreat   ✓ |

Description

| Antimalware Threat Response |

Subscription ⓘ

| Microsoft Azure MVP   ∨ |

Resource group * ⓘ

| Automation   ∨ |

Trigger conditions ⓘ
Choose the trigger conditions that will automatically trigger the configured action.

Select Security Center data types *

| Threat detection alerts   ∨ |

Alert name contains ⓘ

| Antimalware   ✓ |

Alert severity *

| Medium, High   ∨ |

Actions
Configure the Logic App that will be triggered.
Choose an existing Logic App or visit the Logic Apps page to create a new one

Show Logic App instances from the following subscriptions *

| 7 selected   ∨ |

Logic App name ⓘ

| ASC-ThreatDetectionResponse (Security Center alerts connector) ∨ |
Refresh   View logic app

[Create]   [Cancel]

Figure 3.8 – Add workflow automation – threat detection alerts

11. From the **Logic App name** menu, select a Logic App that will be triggered.

12. The menu contains Logic Apps based on the choices you made in this recipe, in *Step 10*. In this example, I selected a Logic App that will send an email when a threat is triggered in Microsoft Defender for Cloud. See the *Creating Logic Apps for use in Microsoft Defender for Cloud* recipe of this chapter for instructions on how to create a Logic App that can be used for Microsoft Defender for Cloud automation.

13. Click **Create**.

14. In the **Microsoft Defender for Cloud Workflow automation** blade, you can identify a new workflow automation entry and check that it is **Enabled**.

Figure 3.9 – Microsoft Defender for Cloud Workflow automation – threat detection response

How it works...

To configure Microsoft Defender for Cloud workflows, you must have the appropriate Microsoft Defender for Cloud and Logic App permissions. You must be a **Security admin** or **Owner** on the resource group, must have write permissions for the target resource, must have **Logic App Creator** permission to run or execute existing Logic Apps, and **Logic App Contributor** permission to create or modify a Logic App.

Additionally, you must have the appropriate credentials to sign into a Logic App connector, if required.

When configuring automation workflows, you can consult the **Security alerts – a reference guide** page for a list of security alerts supported in **Microsoft Defender for Cloud**: https://docs.microsoft.com/en-us/azure/defender-for-cloud/alerts-reference.

Automating Microsoft Defender for Cloud recommendation responses

Comparable to the Threat Detection Alert Response automation workflow, you can automate Microsoft Defender for Cloud recommendation responses.

In this recipe, you will learn how to automate Microsoft Defender for Cloud recommendation responses.

This example will show you how to automate a recommendation response to **Multi-Factor Authentication (MFA)** recommendations.

Getting ready

Open a web browser and navigate to `https://portal.azure.com`.

How to do it...

To automate a recommendation alert response in Defender for Cloud, complete the following steps:

1. In the Azure portal, open **Microsoft Defender for Cloud**.
2. From the left menu, select **Workflow automation**.
3. From the top menu, click **+ Add workflow automation**. An **Add workflow automation** window will open on the right that has three sections: **General**, **Trigger conditions**, and **Action**.
4. Under the **General** section, in the **Name** field, type in a name without spaces; for example, `MFA`.
5. In the **Description** field, type in a description of a workflow's automation; for example, `MFA Recommendations response`.
6. Select a subscription and a resource group.

7. Under the **Trigger conditions** section, from the **Defender for Cloud data type** menu, select **Recommendation**.

> **Note**
>
> Before filtering any recommendations, notice the **Recommendation severity** menu. It contains **Low**, **Medium**, and **High** severities, all of which can be individually selected. **Recommendation name** filtering and **Recommendation severity** are mutually exclusive. After you apply a **Recommendation name** filter in the next step, the **Recommendation severity** menu will become grayed out.

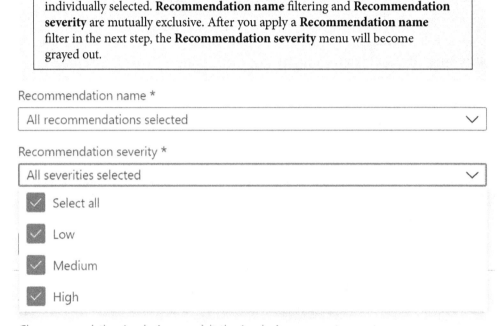

Figure 3.10 – Add workflow automation – Recommendation severity

8. **Recommendation name** filtering and **Recommendation severity** are mutually exclusive. After you apply the **Recommendation name** filter in the next step, the **Recommendation severity** menu will become grayed out.

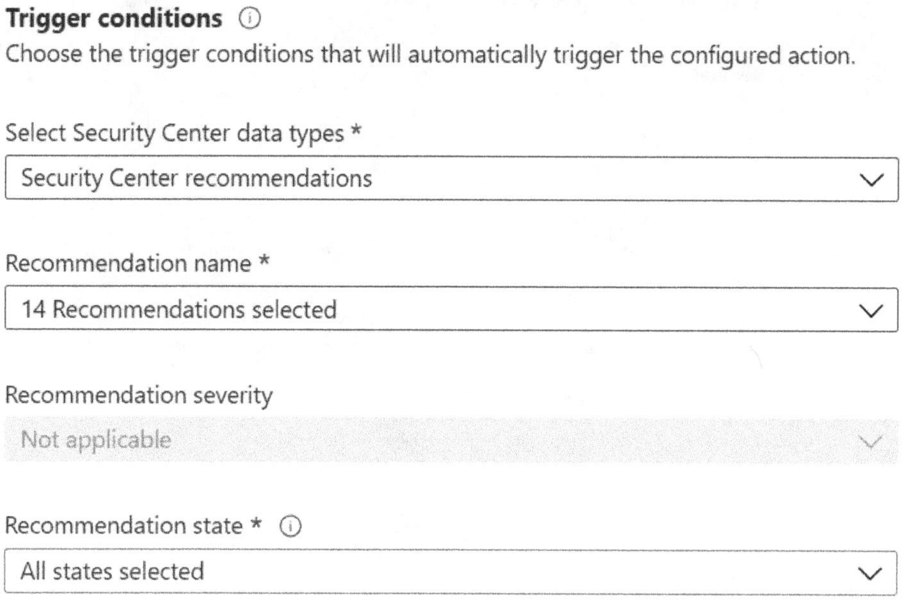

Trigger conditions ⓘ

Choose the trigger conditions that will automatically trigger the configured action.

Select Security Center data types *

Security Center recommendations ⌄

Recommendation name *

14 Recommendations selected ⌄

Recommendation severity

Not applicable ⌄

Recommendation state * ⓘ

All states selected ⌄

Figure 3.11 – Add workflow automation – Recommendation severity not applicable

9. From the **Recommendation name** menu, click **Select all** to deselect
    all recommendations.

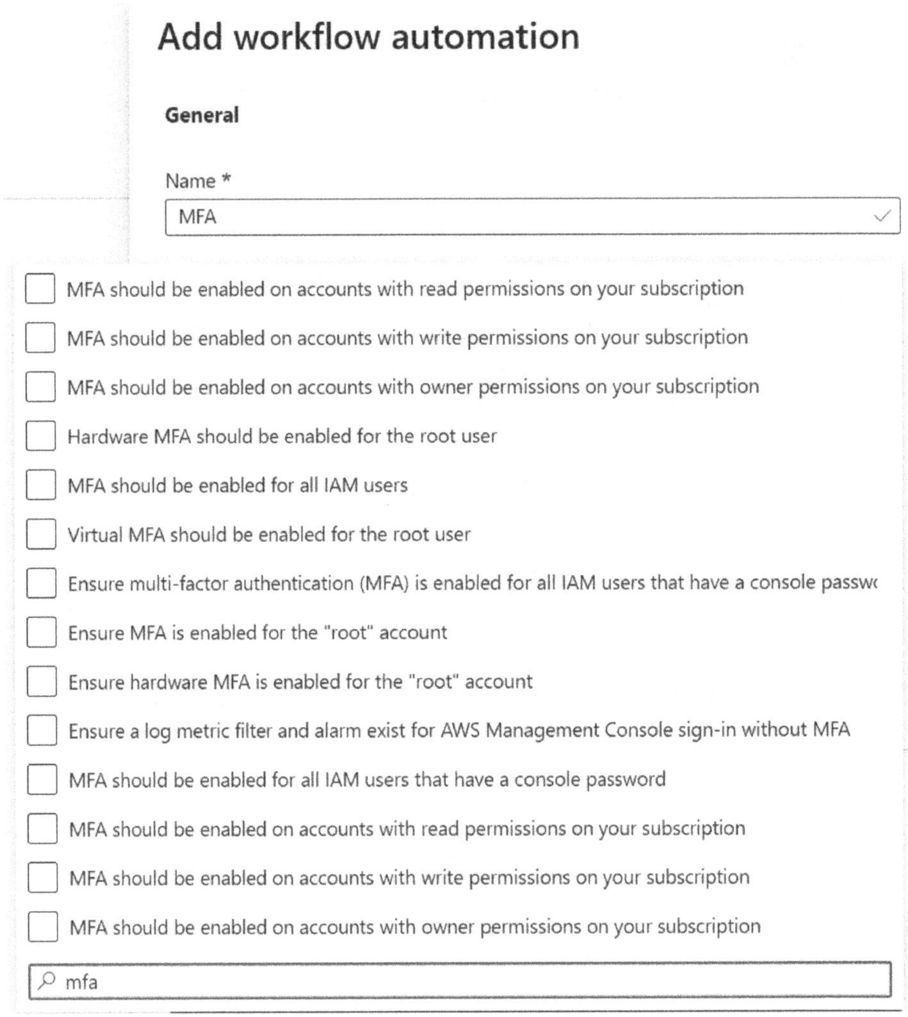

Add workflow automation

General

Name *

| MFA | ✓ |

☐ MFA should be enabled on accounts with read permissions on your subscription

☐ MFA should be enabled on accounts with write permissions on your subscription

☐ MFA should be enabled on accounts with owner permissions on your subscription

☐ Hardware MFA should be enabled for the root user

☐ MFA should be enabled for all IAM users

☐ Virtual MFA should be enabled for the root user

☐ Ensure multi-factor authentication (MFA) is enabled for all IAM users that have a console passwo

☐ Ensure MFA is enabled for the "root" account

☐ Ensure hardware MFA is enabled for the "root" account

☐ Ensure a log metric filter and alarm exist for AWS Management Console sign-in without MFA

☐ MFA should be enabled for all IAM users that have a console password

☐ MFA should be enabled on accounts with read permissions on your subscription

☐ MFA should be enabled on accounts with write permissions on your subscription

☐ MFA should be enabled on accounts with owner permissions on your subscription

🔎 mfa

Figure 3.12 – Add workflow automation – Filtering by recommendation name

10. In the search pane, type MFA to filter for recommendations that contain the
    word **MFA**.

11. Select all the filtered recommendations.

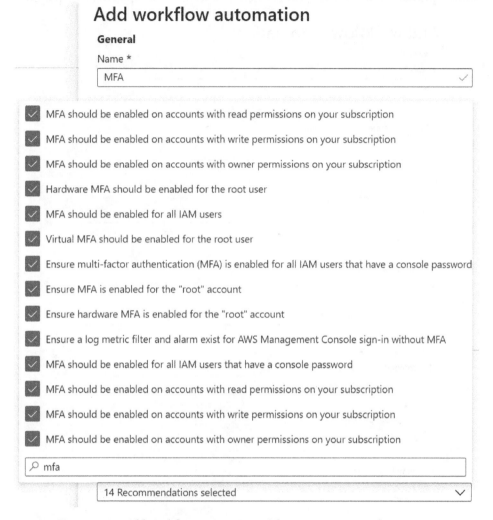

Figure 3.13 – Add workflow automation – Selecting a recommendation name

12. **Recommendation state** contains filters based on a recommendation state: **Healthy**, **Unhealthy**, or **Not applicable**. Leave all of these selected.

13. From the **Actions** section, under **Show Logic App instances from the following subscriptions**, choose subscriptions where you want Logic Apps to be displayed.

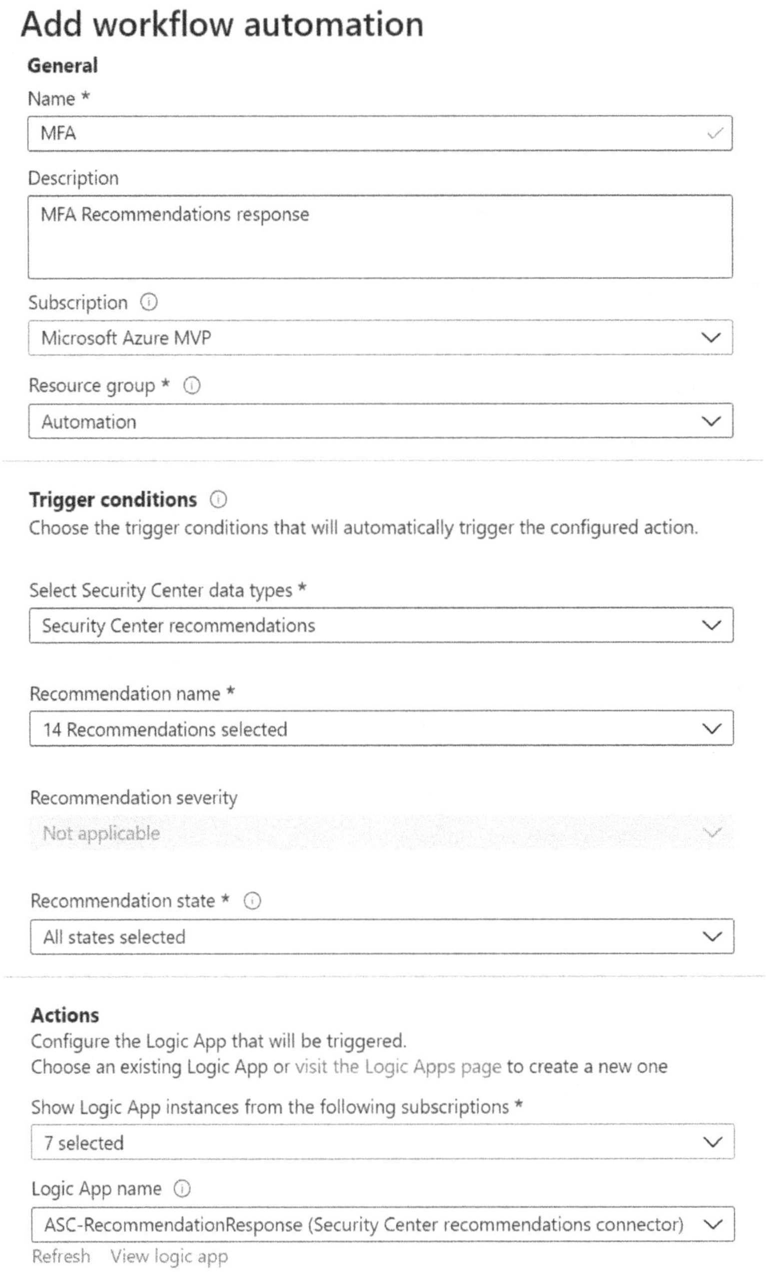

Add workflow automation

General

Name *

| MFA | ✓ |

Description

| MFA Recommendations response |

Subscription ⓘ

| Microsoft Azure MVP | ∨ |

Resource group * ⓘ

| Automation | ∨ |

Trigger conditions ⓘ

Choose the trigger conditions that will automatically trigger the configured action.

Select Security Center data types *

| Security Center recommendations | ∨ |

Recommendation name *

| 14 Recommendations selected | ∨ |

Recommendation severity

| Not applicable | ∨ |

Recommendation state * ⓘ

| All states selected | ∨ |

Actions

Configure the Logic App that will be triggered.
Choose an existing Logic App or visit the Logic Apps page to create a new one

Show Logic App instances from the following subscriptions *

| 7 selected | ∨ |

Logic App name ⓘ

| ASC-RecommendationResponse (Security Center recommendations connector) | ∨ |

Refresh   View logic app

Figure 3.14 – Add workflow automation – Security Center recommendations

14. From the **Logic App name** menu, select a Logic App that will be triggered.

15. Click **Create**.

16. In the **Microsoft Defender for Cloud Workflow automation** blade, you can identify a new workflow automation entry and check that it is **Enabled**.

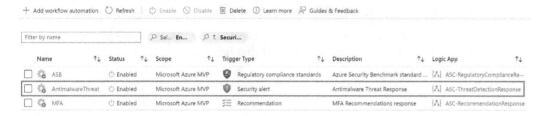

Figure 3.15 – Microsoft Defender for Cloud Workflow automation – Security Center alert trigger

How it works...

When a recommendation state is **Healthy**, Microsoft Defender for Cloud detects it as healthy, and a recommendation no longer applies to the resource. **Unhealthy** state recommendations require your attention to resolve the security recommendation. If a recommendation, for example, was disabled in the security policy, it will have a **Not applicable** state.

Automating regulatory compliance standards responses

Analogous to Threat Detection Alert Response automation and Microsoft Defender for Cloud recommendations response workflow automation, you can trigger automation when a Microsoft Defender for Cloud regulatory compliance assessment is created or triggered.

In this recipe, you will learn how to automate Microsoft Defender for Cloud regulatory compliance standards responses.

This example will show you how to automate a response to changes to the Azure-Security-Benchmark standard.

Getting ready

Open a web browser and navigate to https://portal.azure.com.

How to do it...

To automate a recommendation alert response in Defender for Cloud, complete the following steps:

1. In the Azure portal, open **Microsoft Defender for Cloud**.

2. From the left menu, select **Workflow automation**.

3. From the top menu, click **+ Add workflow automation**. An **Add workflow automation** window will open on the right that has three sections: **General**, **Trigger conditions**, and **Action**.

4. Under the **General** section, in the **Name** field, type in a name without spaces; for example, ASB.

5. In the **Description** field, type in a description of a workflow's automation; for example, Azure Security Benchmark standard changes alert.

6. Select a subscription and a resource group.

7. Under the **Trigger conditions** section, from the **Defender for Cloud data type** menu, select **Regulatory compliance standards**.

8. From the **Compliance standard** menu, click **Select all** to deselect all standards.

9. Select the **Azure-Security-Benchmark** standard.

Figure 3.16 – Add workflow automation – Compliance standard

10. **Compliance control state** contains filters based on the state of the selected compliance controls: **Failed**, **Passed**, **Skipped**, or **Unsupported**. Leave all of them selected.

11. For the **Actions** section, under **Show Logic App instances from the following subscriptions**, choose the subscriptions where you want the Logic Apps to be displayed.

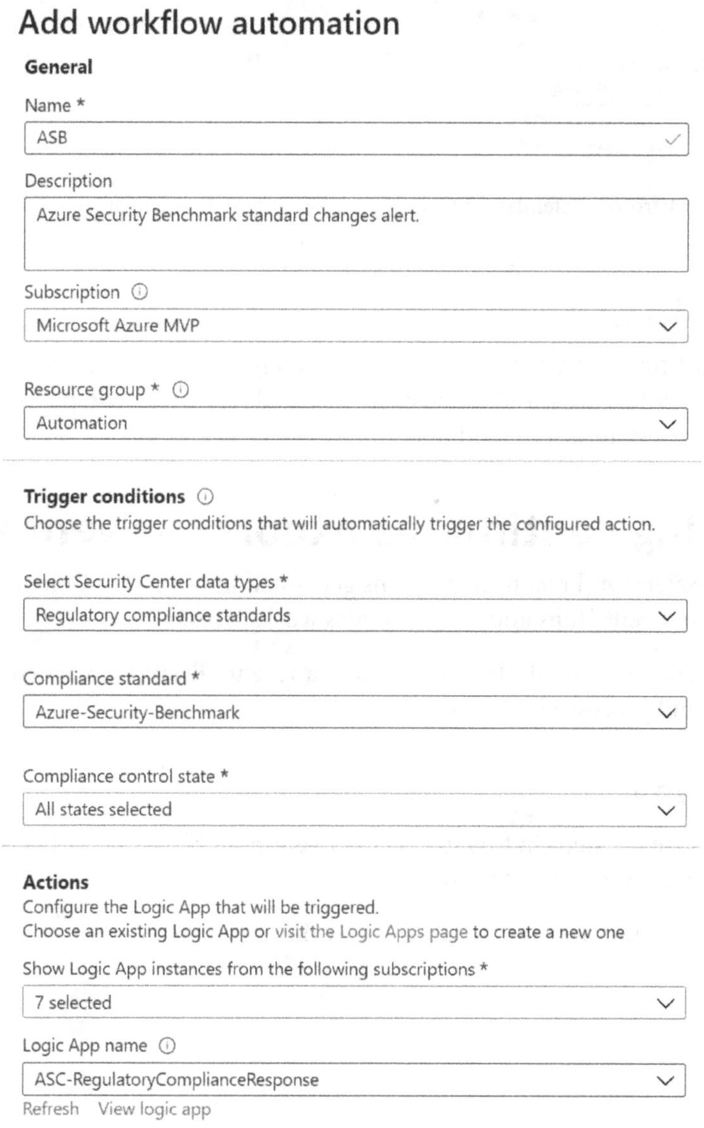

Add workflow automation

General

Name *

ASB

Description

Azure Security Benchmark standard changes alert.

Subscription ⓘ

Microsoft Azure MVP

Resource group * ⓘ

Automation

Trigger conditions ⓘ
Choose the trigger conditions that will automatically trigger the configured action.

Select Security Center data types *

Regulatory compliance standards

Compliance standard *

Azure-Security-Benchmark

Compliance control state *

All states selected

Actions
Configure the Logic App that will be triggered.
Choose an existing Logic App or visit the Logic Apps page to create a new one

Show Logic App instances from the following subscriptions *

7 selected

Logic App name ⓘ

ASC-RegulatoryComplianceResponse

Refresh   View logic app

Figure 3.17 – Add workflow automation – Regulatory compliance standards

12. From the **Logic App name** menu, select a Logic App that will be triggered.

13. Click **Create**.

14. In the **Microsoft Defender for Cloud Workflow automation** blade, you can identify a new workflow automation entry and check that it is **Enabled**.

	Name	↑↓	Status	↑↓	Scope	↑↓	Trigger Type	↑↓	Description	↑↓	Logic App	↑↓
☐	ASB		⏱ Enabled		Microsoft Azure MVP		Regulatory compliance standards		Azure Security Benchmark standard ...		ASC-RegulatoryComplianceRe...	
☐	AntimalwareThreat		⏱ Enabled		Microsoft Azure MVP		Security alert		Antimalware Threat Response		ASC-ThreatDetectionResponse	
☐	MFA		⏱ Enabled		Microsoft Azure MVP		Recommendation		MFA Recommendations response		ASC-RecommendationResponse	

Figure 3.18 – Microsoft Defender for Cloud Workflow automation – Regulatory compliance standards trigger

How it works...

Every time a regulatory compliance standard assessment is completed, which happens approximately every 12 hours, and if one of the selected regulatory compliance standards assessments changes state, a selected Logic App will execute.

Configuring continuous export to Event Hub

Detailed security alerts and recommendations generated in Microsoft Defender for Cloud can be exported to Event Hubs and Log Analytics workspaces.

Specific recommendations or alerts can be sent to an Azure Event Hub, where these events can be analyzed and processed even further.

Getting ready

Before you complete the steps in this recipe, the **Event Hub Namespace**, **Event Hub**, and **Event Hub Policy** resources must be available.

Open a web browser and navigate to https://portal.azure.com.

How to do it...

To configure the continuous export of Defender for Cloud alerts and recommendations data to Event Hub, complete the following steps:

1. In the Azure portal, open **Microsoft Defender for Cloud**.
2. From the left menu, under **Management**, select **Pricing & Settings**.
3. Click on an Azure subscription where you want to configure the data export.
4. From the left menu, select **Continuous export**.
5. From the right blade, click on the **Event hub** tab.
6. The **Export enabled** button allows you to enable or disable a data export. Select **On**.
7. Under **Exported data types**, select the checkboxes for the data types you want to export. Select all the checkboxes.

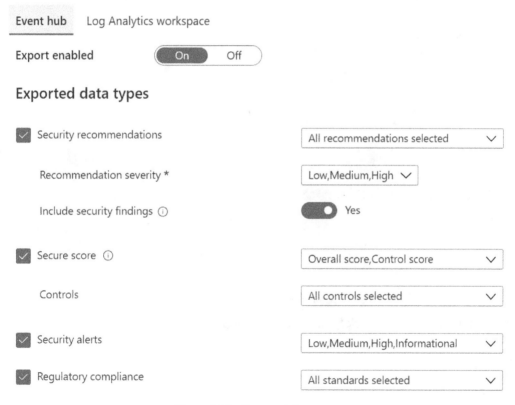

Figure 3.19 – Exported data types

8. Under **Export frequency**, select both checkboxes. **Streaming updates** allows you to export updates in real time, while **Snapshots** allows you to export snapshots of the data types selected under **Exported data types**: regulatory compliance, secure score controls, and the overall security score.

9. Under **Export configuration**, choose a **Resource group** where the export configuration will reside.

10. Under **Export target**, select **Subscription**, **Event Hub namespace**, **Event Hub name**, and **Event Hub policy name**.

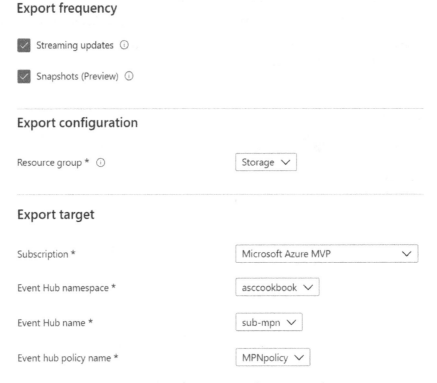

Figure 3.20 – Export frequency, configuration, and target

11. From the top menu, click **Save** to save the configuration.

How it works...

Although the feature is called **Continuous**, it also allows you to configure data exports periodically. You can export various types of data: security alerts, security recommendations, secure scores, security findings, and regulatory compliance data.

Continuous exports can be configured and managed using the Microsoft Defender for Cloud API, as well as deployed at scale using Azure Policy.

Configuring continuous export to a Log Analytics workspace

Specific recommendations or alerts can be sent to a Log Analytics workspace, where these events can be viewed, queried, analyzed, and processed further.

Getting ready

Before you complete the steps in this recipe, you must have a Log Analytics workspace available.

Open a web browser and navigate to `https://portal.azure.com`.

How to do it...

To configure the continuous export of Microsoft Defender for Cloud alerts and recommendations data to a Log Analytics workspace, complete the following steps:

1. In the Azure portal, open **Microsoft Defender for Cloud**.
2. From the left menu, under **Management**, select **Pricing & Settings**.
3. Click on an Azure subscription where you want to configure the data export.
4. From the left menu, select **Continuous export**.
5. From the right blade, click on the **Log Analytics Workspace** tab.
6. The **Export enabled** button allows you to enable or disable a data export. Select **On**.
7. Under **Exported data types**, select the checkboxes for the data types you want to export. Select all the checkboxes.

8. Under **Export frequency**, select both checkboxes. **Streaming updates** allows you to export updates in real time, while **Snapshots** allows you to export snapshots of the data types selected under **Exported data types**: regulatory compliance, secure score controls, and the overall security score.

9. Under **Export configuration**, choose a **Resource group** where the export configuration will reside.

10. Under **Export target**, select a subscription and log analytics workspace where the Microsoft Defender for Cloud data will be exported.

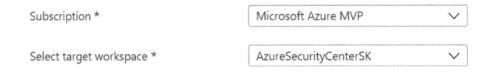

Figure 3.21 – Export frequency, configuration, and target

11. From the top menu, click **Save** to save the configuration.

How it works...

Once the Microsoft Defender for Cloud data is stored in a Log Analytics workspace, you can use **Kusto Query Language** (**KQL**) in Log Analytics for further analysis.

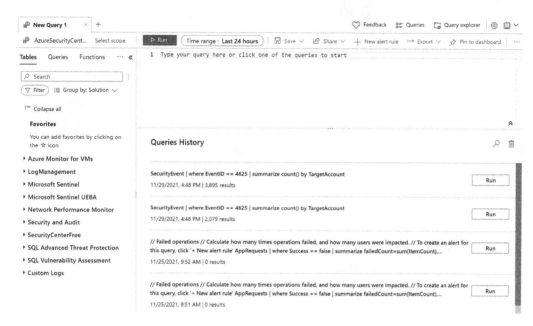

Figure 3.22 – Log Analytics workspace

Security alerts and recommendations are stored in the **SecurityAlert** and **SecurityRecommendation** tables, depending on whether you have Microsoft Defender for Cloud plans enabled – **Security and Audit** and **SecurityCenterFree**.

4

Secure Score and Recommendations

Microsoft Defender for Cloud has two main goals: to help you understand your existing security situation and to help you effectively increase the security of your infrastructure.

Defender for Cloud continuously evaluates your resources, subscriptions for security problems, and aggregates results into a single score. The secure score is shown as a percentage value and it also gives you the possibility to compare the secure score and security level to previous scores and view them as a timeline, revealing the trend of your infrastructure security levels over time.

Recommendations are grouped into categories, or security controls – logical groups of associated security recommendations.

This chapter will help you get the most out of Microsoft Defender for Cloud's two main objectives, and you will learn how to work with and interpret the secure score and manage security recommendations in Microsoft Defender for Cloud.

We will cover the following recipes in this chapter:

- Understanding, filtering, and sorting recommendations
- Downloading a recommendation report
- Creating a recommendation exemption rule
- Creating a recommendation enforcement rule
- Preventing creating resources using a **Deny** rule
- Disabling a recommendation
- Fixing recommendations on affected resources
- Managing a recommendation query in Azure Resource Graph Explorer
- Getting a secure score using Azure Resource Graph

Technical requirements

To successfully complete the recipes in this chapter, the following are required:

- An Azure subscription.
- Microsoft Excel.
- A web browser – preferably Microsoft Edge.
- A Microsoft Defender for Cloud plan.
- A PowerBI Pro account.
- The PowerBI Desktop application.
- Resources in an Azure subscription, such as virtual machines, storage, SQL Server, and Logic Apps. Microsoft Defender for Cloud will create resource recommendations based on available resources.

The code samples can be found at `https://github.com/PacktPublishing/Microsoft-Defender-for-Cloud-Cookbook`.

Understanding, filtering, and sorting recommendations

Microsoft Defender for Cloud uses Azure Policy to define security policy definitions, that is, rules about security conditions that can be grouped in Azure Policy initiatives.

Based on security policies and security initiatives, Microsoft Defender for Cloud determines the compliance status of monitored resources and provides recommendations to correct any potential security misconfigurations.

Depending on the security status and number of monitored resources, the recommendations list can grow significantly. To help you focus your security remediation efforts on controls that matter the most to you, this recipe will show you how to understand the recommendations blade, filter, and sort security recommendations in Microsoft Defender for Cloud.

Getting ready

Open a web browser and navigate to `https://portal.azure.com`.

How to do it...

To filter, group, and sort security recommendations in Microsoft Defender for Cloud, take the following steps:

1. In the Azure portal, open **Microsoft Defender for Cloud**.
2. On the left menu, click **Recommendations**.

3. The **Recommendations** blade has two tabs: **Secure score recommendations** and **All recommendations**:

Figure 4.1 – Microsoft Defender for Cloud recommendations page – Secure score recommendations

4. Click on the **Secure score recommendations** tab if it is not selected.

    The top part of the recommendations blade shows the secure score percentage, **Resource health** status, **Completed controls** status, and **Completed recommendations** status.

    Microsoft Defender for Cloud heavily uses color coding as a visual aid to help quicker and more accurate identification of resources' status, and it is present here as well.

    The bottom part shows a list of recommendations that directly affect the secure score. The recommendation columns include **Controls, Max score, Current Score, Potential Score increase, Unhealthy resources, Resource health**, and **Actions**.

    Additionally, the lower part of the blade contains a search field, filter controls, and a sorting menu.

5. In the middle of the page, on the right side, identify the sorting menu. By default, the recommendations list is sorted by max score. Click on the sorting menu to collapse the menu and select **Sort by potential score increase**:

These recommendations directly affect your secure score. They're grouped into security controls, each representing a risk category. Focus your efforts on controls worth the most points, and fix all recommendations for all resources in a control to get the max points.   Learn more >

| 🔍 Search recommend... | Control status : **All** | Recommendation status : **2 Selected** | Recommendation maturity : **All** | Severity : **All** | Sort by potential scor... ⌄ |
| Expand all | Resource type : **All** | Response actions : **All** | Contains exemptions : **All** | Environment : **All** | Reset filters |

Controls		Max score	Current Score	Potential score increa...	Unhealthy resources	Resource health	Actions
>	Remediate vulnerabilities	6	0.55	+ **9%** (5.45 points)	10 of 11 resources		
>	Secure management ports	8	2.67	+ **9%** (5.33 points)	6 of 9 resources		
>	Enable encryption at rest	4	0.44	+ **6%** (3.56 points)	8 of 14 resources		
>	Remediate security configurations	4	1.09	+ **5%** (2.91 points)	8 of 14 resources		
>	Restrict unauthorized network access	4	1.71	+ **4%** (2.29 points)	8 of 44 resources		
>	Apply adaptive application control	3	0.9	+ **4%** (2.1 points)	7 of 10 resources		
>	Enable MFA	10	8.57	+ **2%** (1.43 points)	1 of 11 resources		

Figure 4.2 – Recommendations – sorted by potential score increase

6.   Click on the **All recommendations** tab:

Figure 4.3 – Microsoft Defender for Cloud recommendations page – All recommendations

7.   At the top of the blade is information about **Completed recommendations (by severity)** and **Resource health**.

This page view has a list of all the recommendations you can use to harden your resources and, for support, the columns' names and information have been adjusted from the previous view: **Recommendation**, **Unhealthy resources**, **Resource health**, **Initiative**, and **Actions**. The last column is useful as it displays the name of the applicable security initiative to which a recommendation belongs.

Accordingly, the filter options are slightly different as well, displaying **Initiative** instead of **Control status**.

8. The column headers show the column names and two arrows, pointing up and down, depicting the sorting status. Click on any column header to sort in alphabetic or numerical order.

9. Multiple clicks on the same column header cycles through the sorting orders, from A to Z or lower to higher number value, or from Z to A or from higher to lower column value. Click repeatedly on the same column header to cycle through the sorting orders:

Figure 4.4 – Recommendations – sorting order

10. To filter the recommendations list, use filters. In this example, you will filter the recommendations based on **Severity** and **Resource type**.

    In the filter list, click on **Severity**. A drop-down **Severity** menu appears:

Figure 4.5 – Filter by Severity

11. From the **Severity** drop-down list, if selected, deselect the **Medium** and **Low** severity statuses and leave only the **High** severity status selected.

12. Click on the **Severity** menu or anywhere on the page to apply and close the **Severity** filter.

13. Click on the **Resource type** filter. A **Resource type** filter dropdown will appear.

14. Click to deselect **Select all**:

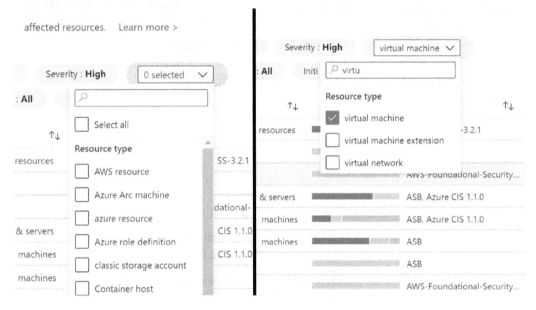

Figure 4.6 – Filter by Resource type

15. In the filter menu search field, type `Virtual`.

16. Click and select the **virtual machine** filter.

17. Click on the **Resource type** menu or anywhere on the page to apply and close the **Resource type** filter:

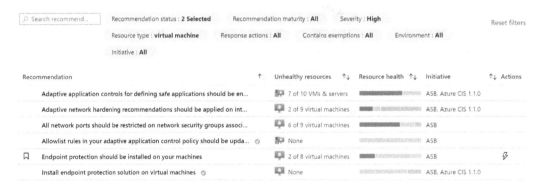

Figure 4.7 – Filtered recommendation list

18. To reset and turn off any applied filters, click **Reset filters**.

How it works...

You can display security recommendations in groups – that is, gathered under a common control name, or a risk category – or display them individually, as a list.

Security recommendations affect your secure score and to work effectively with a large recommendation list, there are several controls and capabilities available: search fields, sorting, grouping, and filtering capabilities.

Downloading a recommendation report

Microsoft Defender for Cloud allows exporting a list of security recommendations, where it can be used for archival or further analysis, for example.

Getting ready

Open a web browser and navigate to `https://portal.azure.com`.

How to do it...

To download a report of security recommendations in Microsoft Defender for Cloud, take the following steps:

1. In the Azure portal, open **Microsoft Defender for Cloud**.
2. On the left menu, click **Recommendations**.

3. At the top of the blade, click **Download CSV report**:

Figure 4.8. – Microsoft Defender for Cloud recommendations – Download CSV report

4. Save the report.

5. Open the report in Microsoft Excel:

Figure 4.9 – Azure security recommendations CSV file

How it works...

Microsoft Defender for Cloud generates a CSV file containing recommendations available and listed at the time of creating a file.

An Azure security list of recommendations in CSV format contains the following fields or columns:

- exportedTimestamp
- subscriptionId
- subscriptionName
- resourceGroup
- resourceType
- resourceName
- resourceId
- recommendationId
- recommendationName
- recommendationDisplayName
- description
- remediationSteps
- severity state
- notApplicableReason
- firstEvaluationDate
- statusChangeDate
- controls
- azurePortalRecommendationLink
- nativeCloudAccountId

Creating a recommendation exemption rule

Microsoft Defender for Cloud allows exempting a recommendation from a list of recommendations in two ways. You can exempt a subscription or management group, or you can exempt a resource. Exempted resources do not impact the secure score.

In this recipe, you will learn how to create an exemption rule for a sample security recommendation.

Getting ready

Open a web browser and navigate to `https://portal.azure.com`.

How to do it...

To create an exemption for a security recommendation in Microsoft Defender for Cloud, take the following steps:

1. In the Azure portal, open **Microsoft Defender for Cloud**.

2. On the left menu, click **Recommendations**.

3. Select the **All recommendations** tab.

4. In the search field, in the middle of the page, to filter the list of recommendations, type `virtual machine`:

Figure 4.10 – Filtered recommendations list

5. From the recommendations list, click on the **Azure Backup should be enabled for virtual machines** recommendation.

6. From the list of **Unhealthy resources**, select a virtual machine:

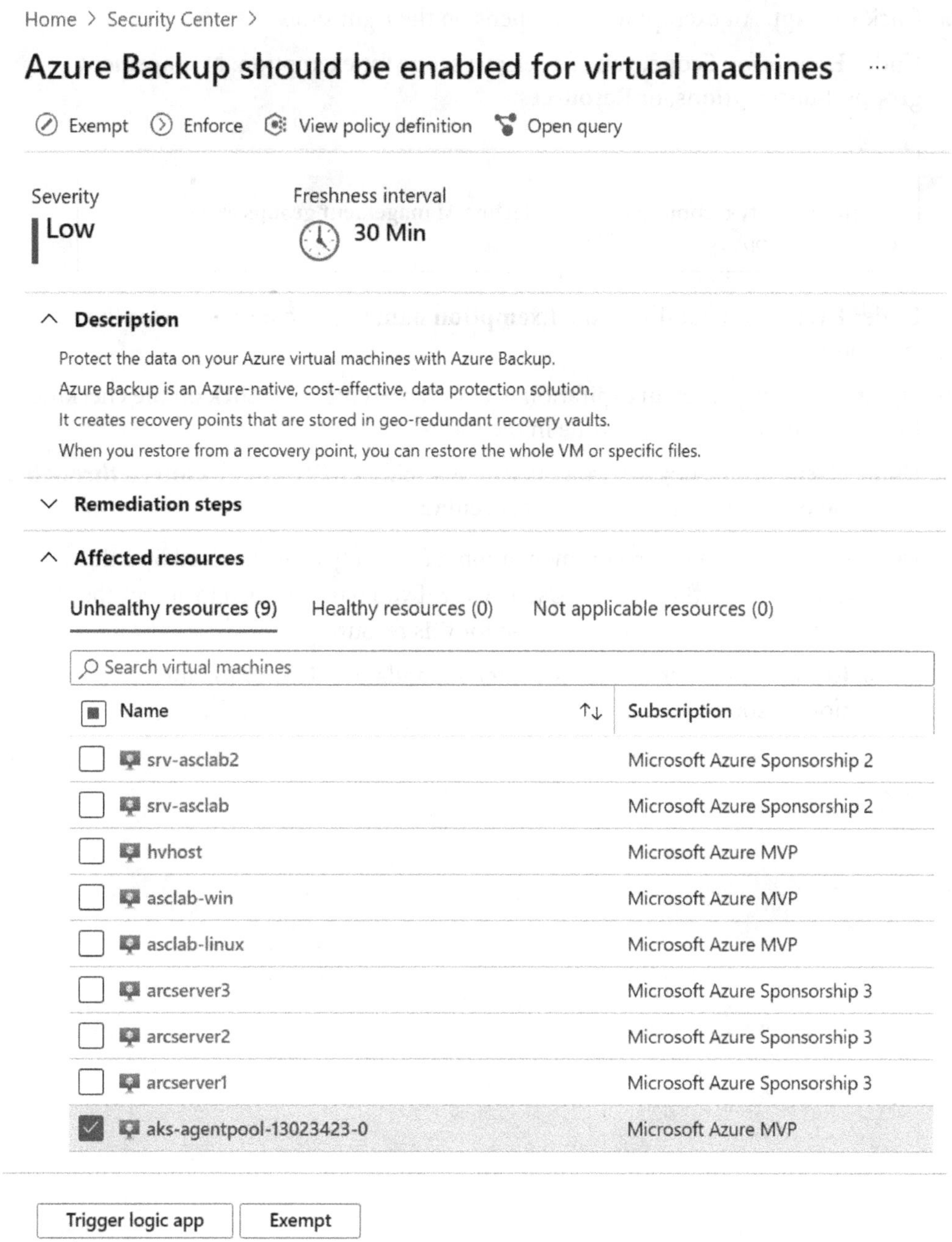

Home > Security Center >

Azure Backup should be enabled for virtual machines   ...

⊘ Exempt   ⊙ Enforce   ⊛ View policy definition   ⚡ Open query

Severity   Freshness interval
| Low   🕐 30 Min

∧ **Description**

Protect the data on your Azure virtual machines with Azure Backup.

Azure Backup is an Azure-native, cost-effective, data protection solution.

It creates recovery points that are stored in geo-redundant recovery vaults.

When you restore from a recovery point, you can restore the whole VM or specific files.

∨ **Remediation steps**

∧ **Affected resources**

Unhealthy resources (9)   Healthy resources (0)   Not applicable resources (0)

🔍 Search virtual machines

■ Name	↑↓	Subscription
☐ 🖥 srv-asclab2		Microsoft Azure Sponsorship 2
☐ 🖥 srv-asclab		Microsoft Azure Sponsorship 2
☐ 🖥 hvhost		Microsoft Azure MVP
☐ 🖥 asclab-win		Microsoft Azure MVP
☐ 🖥 asclab-linux		Microsoft Azure MVP
☐ 🖥 arcserver3		Microsoft Azure Sponsorship 3
☐ 🖥 arcserver2		Microsoft Azure Sponsorship 3
☐ 🖥 arcserver1		Microsoft Azure Sponsorship 3
☑ 🖥 aks-agentpool-13023423-0		Microsoft Azure MVP

[Trigger logic app]   [Exempt]

Figure 4.11 – Microsoft Defender for Cloud recommendation – Azure Backup should be enabled for virtual machines

7. Click **Exempt**. An exempt window opens on the right side.

8. Under **Exemption Scope**, you can select a scope for exemption: **Management groups**, **Subscriptions**, or **Resources**.

> **Note**
>
> For this security recommendation, selecting **Management groups** as an exemption scope is not possible.

9. Under **Exemption details**, in the **Exemption name** field, type a name for the exemption.

10. Optionally, you can set an expiration date for an exemption. Click on the checkbox to set an expiration date and time in the future.

11. Under **Exemption category**, you must choose either **Mitigated (resolved through a third-party service)** or **Waiver (risk acceptance)**.

    Use **Mitigated** when this recommendation has already been resolved by a third-party service or by other means. Use **Waiver** if you have decided to accept the risk of not mitigating this recommendation for this resource.

12. In the **Exemption description** box, enter a mandatory description for the exemption reasons:

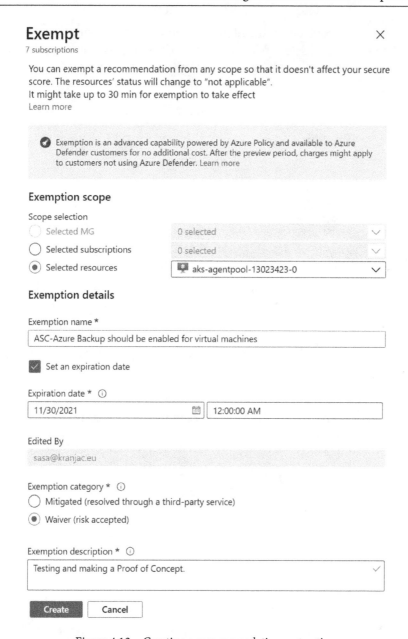

Exempt   ✕
7 subscriptions

You can exempt a recommendation from any scope so that it doesn't affect your secure score. The resources' status will change to "not applicable".
It might take up to 30 min for exemption to take effect
Learn more

> Exemption is an advanced capability powered by Azure Policy and available to Azure Defender customers for no additional cost. After the preview period, charges might apply to customers not using Azure Defender. Learn more

Exemption scope

Scope selection
○ Selected MG | 0 selected   ⌄
○ Selected subscriptions | 0 selected   ⌄
◉ Selected resources | 🖥 aks-agentpool-13023423-0   ⌄

Exemption details

Exemption name *

| ASC-Azure Backup should be enabled for virtual machines |

☑ Set an expiration date

Expiration date * ⓘ

| 11/30/2021   📅 | 12:00:00 AM |

Edited By

sasa@kranjac.eu

Exemption category * ⓘ
○ Mitigated (resolved through a third-party service)
◉ Waiver (risk accepted)

Exemption description * ⓘ

| Testing and making a Proof of Concept.   ✓ |

| Create | | Cancel |

Figure 4.12 – Creating a recommendation exemption

13. Click **Create**. Depending on the freshness interval, policy exemption changes might take several minutes to a few hours to be reflected in Microsoft Defender for Cloud.

14. Close the recommendations blade and return to **Microsoft Defender for Cloud | Recommendations**.

15. To view exempted controls, in the filter, click on **Contains exemptions**. In the menu, deselect **No** and select **Yes**. The exempted controls will show in the list. Observe **Current Score** and **Max score** as both show **Not scored**:

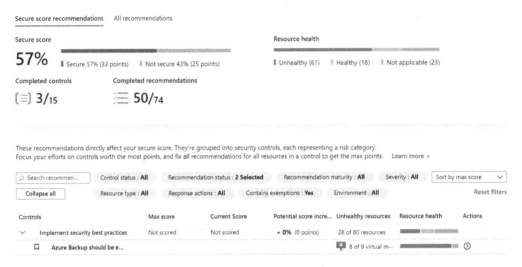

Figure 4.13 – Exempted recommendations and controls

16. Click on the **All recommendations** tab.

17. To view exempted recommendations, in the filter, click on **Contains exemptions**. In the menu, deselect **No** and select **Yes**. The exempted controls and recommendations will show up in the list:

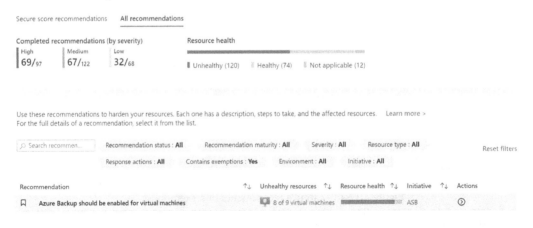

Figure 4.14 – Exempted recommendations

18. Click on the **Azure Backup should be enabled for virtual machines** exempted recommendation:

Home > Security Center >

Azure Backup should be enabled for virtual machines   ⋯   ✕

⊘ Exempt   ⊙ Enforce   ⊡ View policy definition   ⟩ Open query

Severity	Freshness interval	Exempted resources
Low	🕐 **30 Min**	**1** View all exemptions

∧ **Description**

Protect the data on your Azure virtual machines with Azure Backup.

Azure Backup is an Azure-native, cost-effective, data protection solution.

It creates recovery points that are stored in geo-redundant recovery vaults.

When you restore from a recovery point, you can restore the whole VM or specific files.

∨ **Remediation steps**

∧ **Affected resources**

Unhealthy resources (8)   Healthy resources (0)   **Not applicable resources (1)**

🔍 Search virtual machines

Name	↑↓	Subscription	Reason	
🖥 aks-agentpool-13023423-0		Microsoft Azure MVP	Exempt Waiver	⋯

Figure 4.15 – Recommendation details

19. In the middle of the blade, click on the **Not applicable resources** tab. Observe the virtual machine you created an exemption for earlier. Click on the ellipsis (three dots) and select **Manage exemptions**.

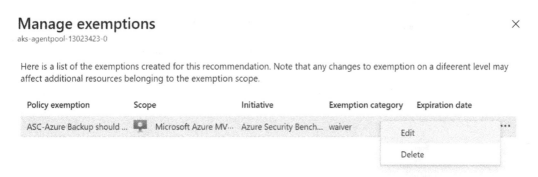

Manage exemptions   ✕
aks-agentpool-13023423-0

Here is a list of the exemptions created for this recommendation. Note that any changes to exemption on a difeerent level may affect additional resources belonging to the exemption scope.

Policy exemption	Scope	Initiative	Exemption category	Expiration date	
ASC-Azure Backup should ...	🖥 Microsoft Azure MV...	Azure Security Bench...	waiver		⋯
			Edit		
			Delete		

Figure 4.16 – Manage exemptions

20. On the **Manage exemptions** blade, you can manage – **Edit** and **Delete** – exemptions created for a particular recommendation.

21. Click **Close**.

22. On the top information strip, select **View all exemptions**:

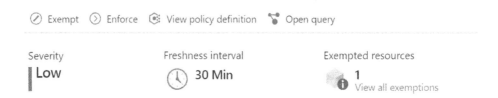

Figure 4.17 – Top information strip on the recommendations blade

23. On the **Exemptions** page, you can view exemptions per **Scope**, **Exemption category**, or **Search** manually by typing a name or ID. In addition, the page shows the number of **total exemptions, exemptions approaching expiration, exemptions expired**, as well as a list of **policy exemptions**:

Figure 4.18 – Exemptions

24. For each policy exemption, click on the ellipsis to access additional policy actions: **edit** or **delete** the exemption, view and edit the assignment, and view the compliance.

How it works...

Any recommendation exemptions will not show and will be hidden on Microsoft Defender for Cloud's recommendations page.

To view specific exempted resources, view the **Not applicable** tab of a recommendation details page.

To view exempted recommendations, apply the **Yes** status to the **Contains exemptions** filter.

Creating a recommendation enforcement rule

Microsoft Defender for Cloud allows enforcing a secure configuration of supported resources. When you use the enforce option, you can prevent security misconfigurations of new resources, considering specific security recommendations.

In this recipe, you will learn how to use the enforce option on non-compliant resources at the time they are created.

Getting ready

To complete the steps in this recipe, you must have a resource that will generate one or more recommendations that currently support the enforce option.

The recommendations that can be used with the enforce option are listed in the *There's more...* section.

Open a web browser and navigate to https://portal.azure.com.

How to do it...

To filter, group, and sort security recommendations in Microsoft Defender for Cloud, take the following steps:

1. In the Azure portal, open **Microsoft Defender for Cloud**.

2. On the left menu, click **Recommendations**.

3. On the **Recommendations** page, select **All recommendations**.

4. To narrow the recommendations list, in the search field, type `logic`:

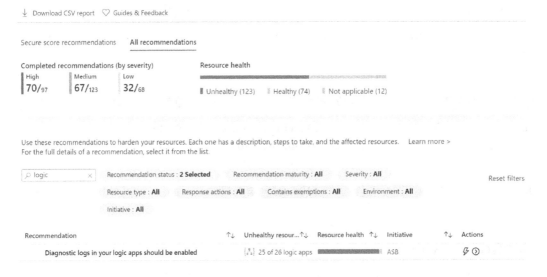

Figure 4.19 – Recommendations list – searching for a recommendation

5. Click on the **Diagnostic logs in your logic apps should be enabled** recommendation.

    Note the two icons in the **Actions** column. The *lightning* icon means a recommendation supports the **Fix** capability to resolve the issue, while the *right arrow* denotes the possibility to **enforce** a security recommendation.

6. At the top of the **Diagnostic logs in your logic apps should be enabled** recommendation page, click on **Enforce**. The **Assign policy** blade opens:

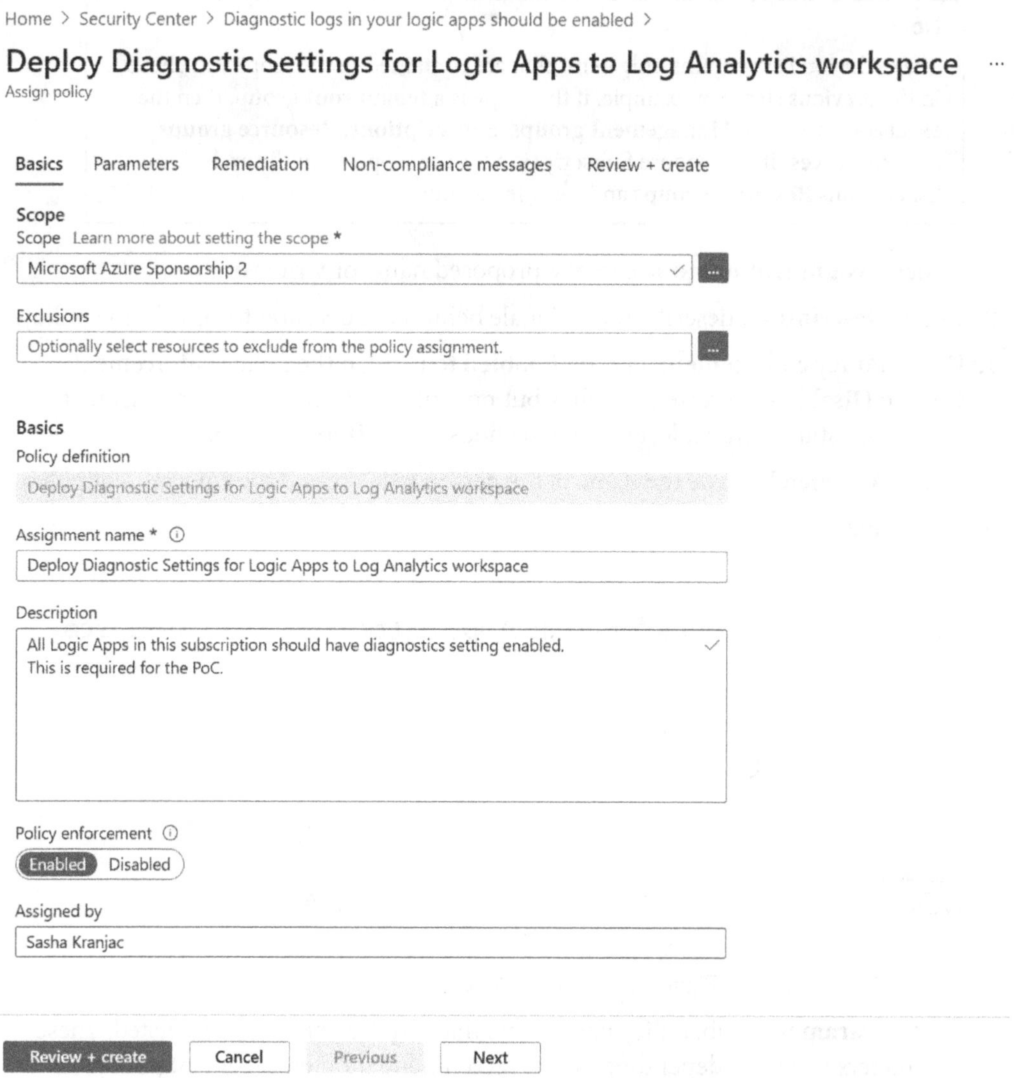

Home > Security Center > Diagnostic logs in your logic apps should be enabled >

Deploy Diagnostic Settings for Logic Apps to Log Analytics workspace   ...

Assign policy

Basics   Parameters   Remediation   Non-compliance messages   Review + create

Scope

Scope   Learn more about setting the scope *

| Microsoft Azure Sponsorship 2 | ✓ | ... |

Exclusions

| Optionally select resources to exclude from the policy assignment. | ... |

Basics

Policy definition

Deploy Diagnostic Settings for Logic Apps to Log Analytics workspace

Assignment name * ⓘ

| Deploy Diagnostic Settings for Logic Apps to Log Analytics workspace |

Description

| All Logic Apps in this subscription should have diagnostics setting enabled. This is required for the PoC. | ✓ |

Policy enforcement ⓘ

(Enabled) Disabled

Assigned by

| Sasha Kranjac |

| Review + create | | Cancel | | Previous | | Next |

Figure 4.20 – Policy assignment – Basics tab

7. Next to the **Scope** field, click on the ellipsis. A window opens, where you can choose a scope for a policy: **Management group**, **Subscription**, or **Resource group**. Select **Subscription** and click **Select**.

8. Next to the **Exclusions** field, click on the ellipsis. A window opens where you can optionally select resources to exclude from the policy assignment.

> **Note**
>
> The exclusion list contains only items that are contained in the scope you chose in the previous step. For example, if the scope is a **tenant root group**, then the exclusions listed are **Management groups**, **Subscriptions**, **Resource groups**, and **Resources**. If you choose **Subscription** for the scope, then the exclusion list contains **Resource groups** and **Resources** only.

9. Under **Assignment name**, accept the proposed name or write your own.

10. Under **Description**, describe the rationale behind the decision to enforce the policy.

11. Under **Policy enforcement**, choose **Enabled** to turn on the policy enforcement. Choose **Disabled** to create the policy but not enforce it. Compliance assessment results will still be available, even if you choose the **Disabled** option.

12. Under **Assigned by**, type the name of the person who created the enforcement policy.

13. Click **Next**.

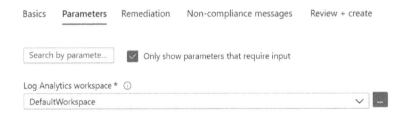

Figure 4.21 – Policy parameters settings

14. On the **Parameters** tab, policy parameters that require your input are listed. These parameters will vary, depending on the recommendation enforcement policy. The **Only show parameters that require input** setting is checked by default, helping you clear the list of settings that do not require your attention.

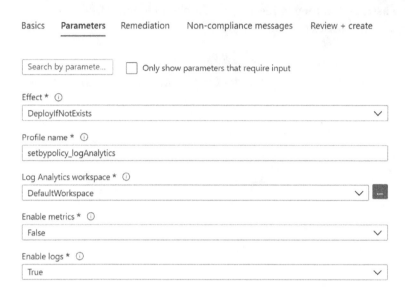

Home > Security Center > Diagnostic logs in your logic apps should be enabled >

Deploy Diagnostic Settings for Logic Apps to Log Analytics workspace   ...
Assign policy

Basics   **Parameters**   Remediation   Non-compliance messages   Review + create

| Search by paramete... | | Only show parameters that require input |

Effect * ⓘ

| DeployIfNotExists | ∨ |

Profile name * ⓘ

| setbypolicy_logAnalytics |

Log Analytics workspace * ⓘ

| DefaultWorkspace | ∨ | ... |

Enable metrics * ⓘ

| False | ∨ |

Enable logs * ⓘ

| True | ∨ |

Figure 4.22 – Editing policy assignment settings

15. For this enforcement, you must choose a Log Analytics workspace, where Logic Apps will send diagnostic information. Choose a workspace from the **Log Analytics workspace** pull-down menu.

16. Click **Next**.

17. At the **Remediation** page, you can check the **Create a remediation task** checkbox to create a policy that will apply and take effect on existing resources. After you select the checkbox, you will have to select a remediation policy from a menu.

18. The **Managed Identity** setting is needed to enable the policy to deploy resources and edit tags on remediated resources. You can choose between **System assigned managed identity** and **User assigned managed identity**. You must specify a location for **System assigned managed identity** or choose an identity from the **Existing user assigned identities** list and its **scope:**

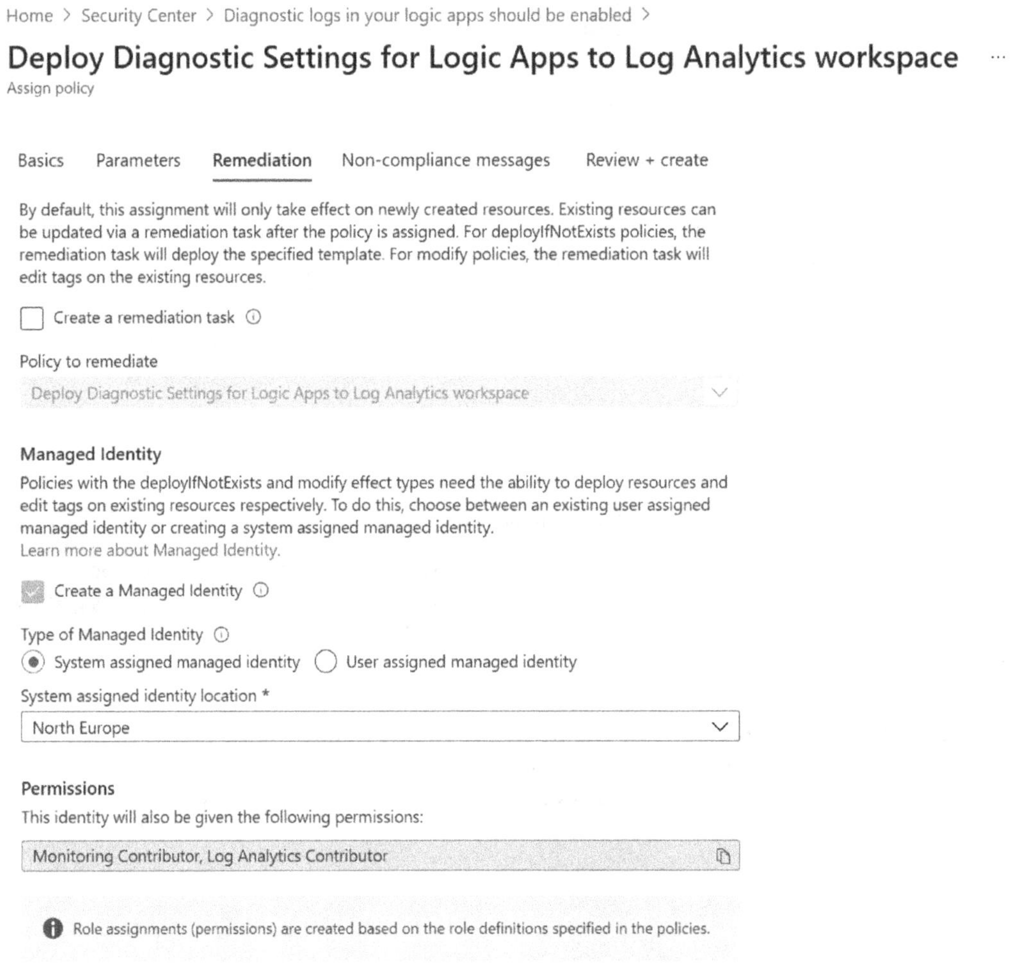

Home > Security Center > Diagnostic logs in your logic apps should be enabled >

Deploy Diagnostic Settings for Logic Apps to Log Analytics workspace   ···
Assign policy

Basics   Parameters   **Remediation**   Non-compliance messages   Review + create

By default, this assignment will only take effect on newly created resources. Existing resources can be updated via a remediation task after the policy is assigned. For deployIfNotExists policies, the remediation task will deploy the specified template. For modify policies, the remediation task will edit tags on the existing resources.

☐ Create a remediation task ⓘ

Policy to remediate

Deploy Diagnostic Settings for Logic Apps to Log Analytics workspace   ∨

Managed Identity

Policies with the deployIfNotExists and modify effect types need the ability to deploy resources and edit tags on existing resources respectively. To do this, choose between an existing user assigned managed identity or creating a system assigned managed identity.
Learn more about Managed Identity.

☑ Create a Managed Identity ⓘ

Type of Managed Identity ⓘ
◉ System assigned managed identity   ◯ User assigned managed identity

System assigned identity location *

North Europe   ∨

Permissions
This identity will also be given the following permissions:

Monitoring Contributor, Log Analytics Contributor   ▢

ⓘ Role assignments (permissions) are created based on the role definitions specified in the policies.

Figure 4.23 – Policy remediation settings

19. Click **Next**.

20. On the **Non-compliance messages** page, enter text that explains to users why a resource is not compliant. The message will be visible in non-compliant resource details or if a resource is denied:

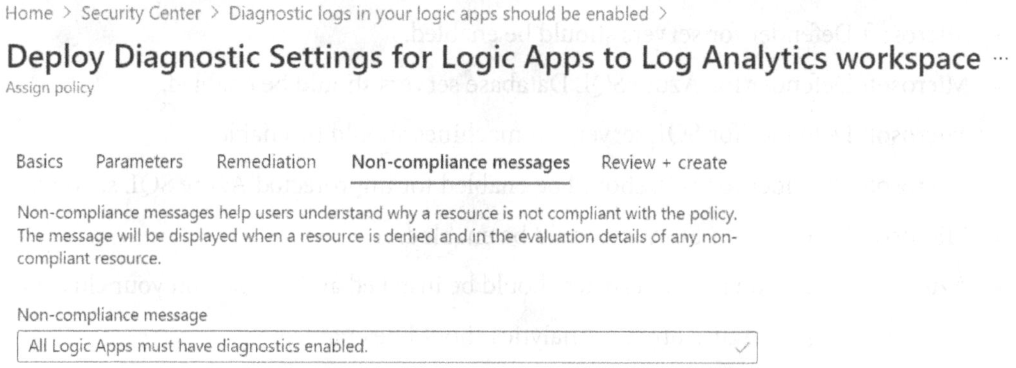

Home > Security Center > Diagnostic logs in your logic apps should be enabled >

Deploy Diagnostic Settings for Logic Apps to Log Analytics workspace ···
Assign policy

Basics   Parameters   Remediation   **Non-compliance messages**   Review + create

Non-compliance messages help users understand why a resource is not compliant with the policy. The message will be displayed when a resource is denied and in the evaluation details of any non-compliant resource.

Non-compliance message

All Logic Apps must have diagnostics enabled.   ✓

Figure 4.24 – Non-compliance messages policy settings

21. Click **Next**.

22. On the **Review** screen, review the settings and click **Create**.

How it works...

To use the **Enforce** option, Azure security policies use the `DeployIfNotExist` effect in Azure Policy to apply settings to newly created and existing resources. For `DeployIfNotExist` policies, the remediation task will deploy the specified template, and to modify policies, the remediation task will edit tags on the existing resources.

There's more...

At this time, the recommendations that can be used with the **Enforce** option are the following:

- Auditing on SQL Server should be enabled.
- Azure Arc-enabled Kubernetes clusters should have the Microsoft Defender for Cloud plan's extension installed.
- Azure Backup should be enabled for virtual machines.
- Microsoft Defender for App Service should be enabled.
- Microsoft Defender for container registries should be enabled.
- Microsoft Defender for DNS should be enabled.

- Microsoft Defender for Key Vault should be enabled.
- Microsoft Defender for Kubernetes should be enabled.
- Microsoft Defender for Resource Manager should be enabled.
- Microsoft Defender for servers should be enabled.
- Microsoft Defender for Azure SQL Database servers should be enabled.
- Microsoft Defender for SQL servers on machines should be enabled.
- Microsoft Defender for SQL should be enabled for unprotected Azure SQL servers.
- Microsoft Defender for Storage should be enabled.
- Azure Policy Add-on for Kubernetes should be installed and enabled on your clusters.
- Diagnostic logs in Azure Stream Analytics should be enabled.
- Diagnostic logs in Batch accounts should be enabled.
- Diagnostic logs in Data Lake Analytics should be enabled.
- Diagnostic logs in Event Hubs should be enabled.
- Diagnostic logs in Key Vault should be enabled.
- Diagnostic logs in Logic Apps should be enabled.
- Diagnostic logs in Search services should be enabled.
- Diagnostic logs in Service Bus should be enabled.
- Preventing creating resources using the **Deny** rule.

Preventing creating resources using a Deny rule

Microsoft Defender for Cloud can prevent creating unhealthy and potentially insecure resources using the **Deny** effect of Azure Policy.

In this recipe, you will learn how to use the **Deny** effect of Azure Policy to enforce infrastructure security by preventing the creation of insecure and non-compliant resources.

Getting ready

To complete the steps in this recipe, you must have a resource that will generate one or more recommendations that currently support the enforce option.

The recommendations that can be used with the enforce option are listed in the *There's more...* section.

Open a web browser and navigate to `https://portal.azure.com`.

How to do it...

To use the **Deny** effect of Azure Policy in Microsoft Defender for Cloud, take the following steps:

1. In the Azure portal, open **Microsoft Defender for Cloud**.

2. On the left menu, click **Recommendations**.

3. On the **Recommendations** page, select **All recommendations**.

4. To narrow the recommendations list, in the search field, type `storage`:

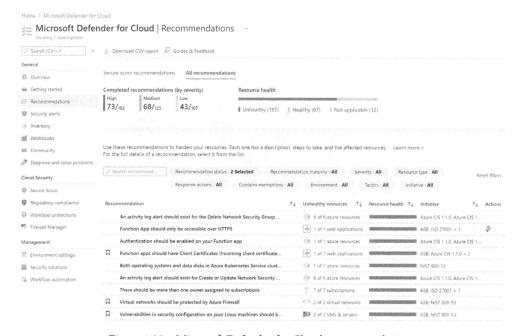

Figure 4.25 – Microsoft Defender for Cloud recommendations

5. Click on the **Storage account public access should be disallowed** recommendation.

    Note the two icons in the **Actions** column. The *lightning* icon means a recommendation supports the **Fix** capability to resolve the issue, while the *dash* or *minus* denotes the possibility to apply the **Deny** action to a security recommendation.

6. At the top of the **Storage account public access should be disallowed** recommendation page, click on **Deny**. A **Deny** blade opens:

Deny - Prevent resource creation   ✕

7 subscriptions

Set the scope for the deny effect of your Azure Policy. The deny effect prevents the creation of resources that don't satisfy the recommendation.
Learn more about the Azure Policy deny effect.

Item	Current status	More
⌄ ☐ [ꞗ] Tenant Root Group (7 of 8 subscriptions)		
⌄ ☐ [ꞗ] MVP (5 of 5 subscriptions)		
☐ [ꞗ] Test (0 of 0 subscriptions)		
☐ 🔑 Courses	Audit	
☐ 🔑 Microsoft Azure MVP	Audit	
☐ 🔑 Microsoft Azure Sponsorship 1	Audit	
☑ 🔑 Microsoft Azure Sponsorship 2	Audit	
☐ 🔑 Microsoft Azure Sponsorship 3	Audit	
☐ [ꞗ] Other (0 of 0 subscriptions)		
⌄ ☐ [ꞗ] Partner (2 of 2 subscriptions)		
☐ 🔑 Microsoft Partner Network	Audit	
☐ 🔑 Visual Studio Enterprise	Audit	

Change to Deny

Figure 4.26 – Deny - Prevent resource creation policy settings

7. In the **Deny – Prevent resource creation** window, select subscriptions to apply the **Deny** effect to. Click **Change to Deny**. The current status of the subscription changes to **Deny**.

8. If you want to revert changes for a subscription, select the ellipsis, then click on the **Change to audit** setting:

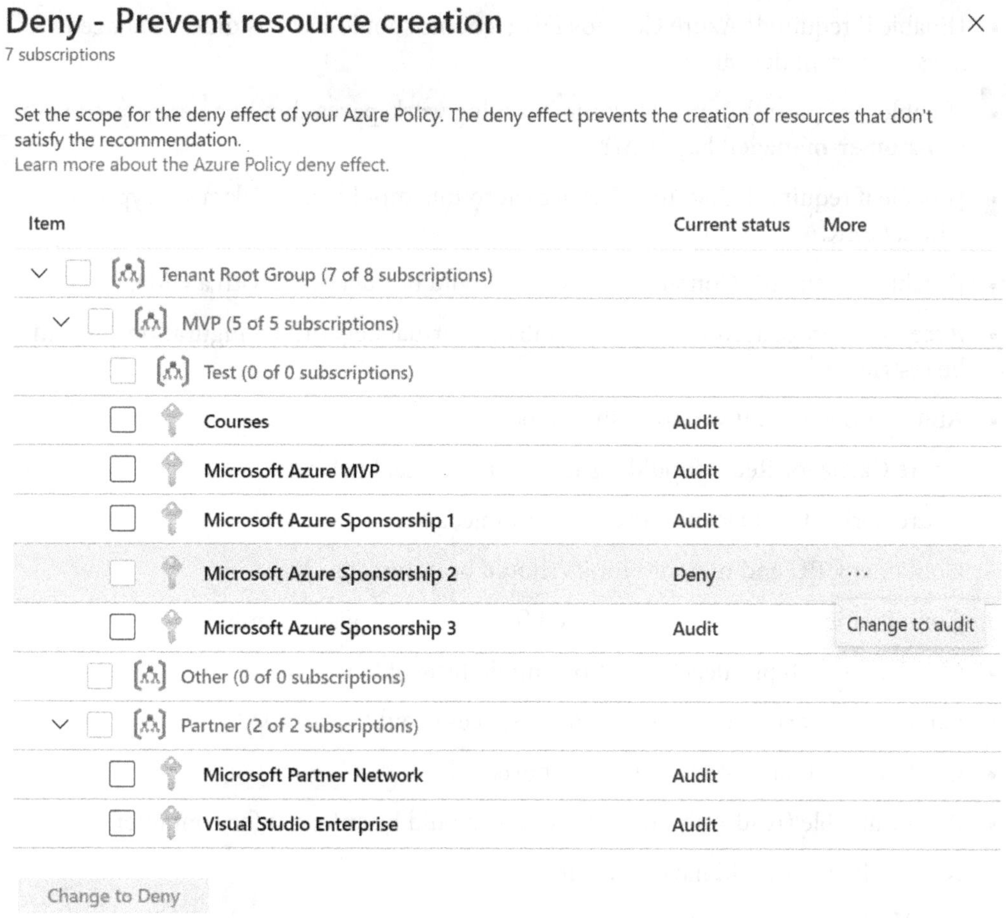

Deny - Prevent resource creation

7 subscriptions

Set the scope for the deny effect of your Azure Policy. The deny effect prevents the creation of resources that don't satisfy the recommendation.
Learn more about the Azure Policy deny effect.

Item	Current status	More
⌄ ☐ [⚙] Tenant Root Group (7 of 8 subscriptions)		
⌄ ☐ [⚙] MVP (5 of 5 subscriptions)		
☐ [⚙] Test (0 of 0 subscriptions)		
☐ 🔑 Courses	Audit	
☐ 🔑 Microsoft Azure MVP	Audit	
☐ 🔑 Microsoft Azure Sponsorship 1	Audit	
☐ 🔑 Microsoft Azure Sponsorship 2	Deny	...
☐ 🔑 Microsoft Azure Sponsorship 3	Audit	Change to audit
☐ [⚙] Other (0 of 0 subscriptions)		
⌄ ☐ [⚙] Partner (2 of 2 subscriptions)		
☐ 🔑 Microsoft Partner Network	Audit	
☐ 🔑 Visual Studio Enterprise	Audit	

Change to Deny

Figure 4.27 – Changing policy settings

9. The current status of the subscription changes to **Audit**.

How it works...

In Azure Policy, **Deny** is used to prevent creating a resource that does not match standards defined through a policy definition.

When creating or updating a resource, **Deny** prevents the request and the request is returned with a 403 (Forbidden) status.

There's more...

At this time, the recommendations that can be used with the **Deny** option are as follows:

- [Enable if required] Azure Cosmos DB accounts should use customer-managed keys to encrypt data at rest.
- [Enable if required] Azure Machine Learning workspaces should be encrypted with a **customer-managed key** (**CMK**).
- [Enable if required] Cognitive Services accounts should enable data encryption with a CMK.
- [Enable if required] Container registries should be encrypted with a CMK.
- Access to storage accounts with firewall and virtual network configurations should be restricted.
- Automation account variables should be encrypted.
- Azure Cache for Redis should reside within a virtual network.
- Azure Spring Cloud should use network injection.
- Container CPU and memory limits should be enforced.
- Container images should be deployed from trusted registries only.
- Containers with privilege escalation should be avoided.
- Containers sharing sensitive host namespaces should be avoided.
- Containers should listen on allowed ports only.
- An immutable (read-only) root filesystem should be enforced for containers.
- Key Vault keys should have an expiration date.
- Key Vault secrets should have an expiration date.
- Key vaults should have purge protection enabled.
- Key vaults should have soft delete enabled.
- Least privileged Linux capabilities should be enforced for containers.
- Only secure connections to your Redis Cache should be enabled.
- The overriding or disabling of containers' AppArmor profiles should be restricted.

- Privileged containers should be avoided.

- Running containers as root user should be avoided.

- Secure transfer to storage accounts should be enabled.

- Service Fabric clusters should have the `ClusterProtectionLevel` property set to `EncryptAndSign`.

- Service Fabric clusters should only use Azure Active Directory for client authentication.

- Services should listen on allowed ports only.

- Storage account public access should be disallowed.

- Storage accounts should be migrated to new Azure Resource Manager resources.

- Storage accounts should restrict network access using virtual network rules.

- Usage of host networking and ports should be restricted.

- Usage of pod `HostPath` volume mounts should be restricted to a known list to restrict node access from compromised containers.

- The validity period of certificates stored in Azure Key Vault should not exceed 12 months.

- Virtual machines should be migrated to new Azure Resource Manager resources.

- A **Web Application Firewall** (**WAF**) should be enabled for Application Gateway.

- A WAF should be enabled for the Azure Front Door service.

Disabling a recommendation

In some cases, if a recommendation is not applicable to your environment, you can prevent it from appearing in the list of security recommendations.

In this recipe, you will learn how to disable a recommendation.

Getting ready

Open a web browser and navigate to `https://portal.azure.com`.

How to do it...

To disable a recommendation and prevent it from appearing in the Microsoft Defender for Cloud recommendation list, take the following steps:

1. In the Azure portal, open **Microsoft Defender for Cloud**.

2. On the left menu, under the **Management** section, click **Environment Settings**.

3. On the **Environment Settings** blade, select a management group or a subscription. For this example, select **Azure subscription**.

4. On the left menu, click on **Security policy**:

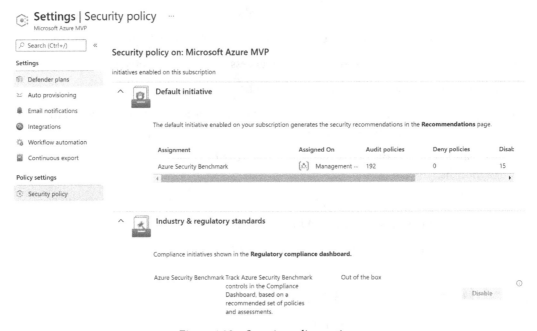

Figure 4.28 – Security policy settings

5. Under **Default initiative**, select the Microsoft Defender for Cloud default policy or your custom initiative policy to disable a recommendation. Click on the ellipsis next to **Microsoft Defender for Cloud default policy** (**Azure Security Benchmark** in *Figure 4.28*) and choose **Edit assignment**. The **Edit Initiative Assignment** blade for the chosen policy opens.

6. Click the **Parameters** tab.

7. Deselect the checkbox for the **Only show parameters that require input** or review option. Displaying all the parameters will take a few moments, as there are a lot of parameters in the policy.

8. From the list, search for the policy parameter you want to disable. To filter the list and to search for a parameter, you can use the search field. For example, type `logic` in the search field.

9. From the list, change the **AuditIfNotExists** setting to **Disabled**:

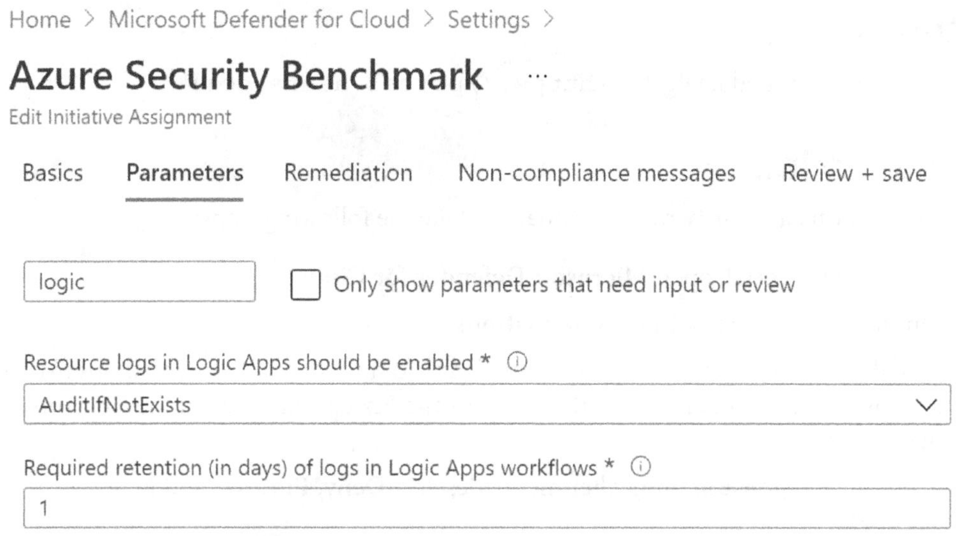

Home > Microsoft Defender for Cloud > Settings >

Azure Security Benchmark   ...
Edit Initiative Assignment

Basics   **Parameters**   Remediation   Non-compliance messages   Review + save

| logic | ☐ Only show parameters that need input or review

Resource logs in Logic Apps should be enabled * ⓘ

| AuditIfNotExists   ⌄ |

Required retention (in days) of logs in Logic Apps workflows * ⓘ

| 1 |

Figure 4.29 – Initiative settings – Parameters tab

10. Click **Review + save**.

How it works...

Microsoft Defender for Cloud uses **Azure Policy** to define security policy definitions, that is, rules about security conditions that can be grouped in Azure Policy initiatives. These policies are then used to generate security recommendations. To prevent displaying security recommendations that are not relevant, you can disable a policy definition parameter that generates the recommendation.

Fixing recommendations on affected resources

Microsoft Defender for Cloud gives you security recommendations or advice and suggestions on how to secure your resources.

In this recipe, you will learn how to apply remediation steps or to fix a resource based on a recommendation.

Getting ready

Open a web browser and navigate to `https://portal.azure.com`.

How to do it...

To remediate or fix a security recommendation, take the following steps:

1. In the Azure portal, open **Microsoft Defender for Cloud**.

2. On the left menu, click **Recommendations**.

3. Not all recommendations support a quick fix. To filter the recommendations list and display only recommendations that support the **Fix** option, click on the **Response actions** filter.

4. From the **Response actions** filter menu, deselect **Deny**, **Enforce**, and **None**, and select **Fix**:

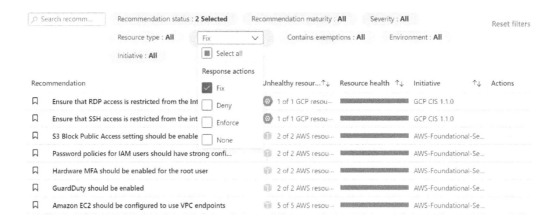

Figure 4.30 – Response actions filter

5. Click anywhere in the blade to close the filter menu and apply the filter. Now you should see the list of recommendations that support the **Fix** option:

> **Note**
>
> The *lightning* icon represents a recommendation that supports the **Fix** capability to resolve the issue, the *dash* or *minus* represents the possibility to apply the **Deny** action to a security recommendation, and the *right arrow* denotes the possibility to **enforce** a security recommendation.

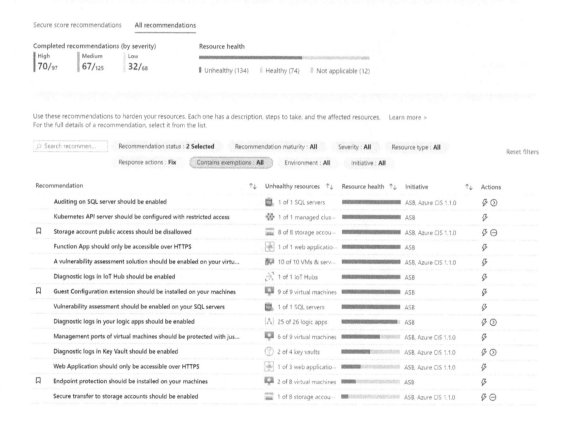

Secure score recommendations   All recommendations

Completed recommendations (by severity)   Resource health

High	Medium	Low
70/97	**67**/125	**32**/68

Unhealthy (134)   Healthy (74)   Not applicable (12)

Use these recommendations to harden your resources. Each one has a description, steps to take, and the affected resources.   Learn more >
For the full details of a recommendation, select it from the list.

Search recommen...   Recommendation status : **2 Selected**   Recommendation maturity : **All**   Severity : **All**   Resource type : **All**   Reset filters

Response actions : **Fix**   Contains exemptions : **All**   Environment : **All**   Initiative : **All**

Recommendation	↑↓	Unhealthy resources ↑↓	Resource health ↑↓	Initiative	↑↓	Actions
Auditing on SQL server should be enabled		1 of 1 SQL servers		ASB, Azure CIS 1.1.0		⚡ ⊙
Kubernetes API server should be configured with restricted access		1 of 1 managed clus···		ASB		⚡
⊓ Storage account public access should be disallowed		8 of 8 storage accou···		ASB, Azure CIS 1.1.0		⚡ ⊖
Function App should only be accessible over HTTPS		1 of 1 web applicatio···		ASB		⚡
A vulnerability assessment solution should be enabled on your virtu...		10 of 10 VMs & serv···		ASB, Azure CIS 1.1.0		⚡
Diagnostic logs in IoT Hub should be enabled		1 of 1 IoT Hubs		ASB		⚡
⊓ Guest Configuration extension should be installed on your machines		9 of 9 virtual machines		ASB		⚡
Vulnerability assessment should be enabled on your SQL servers		1 of 1 SQL servers		ASB		⚡
Diagnostic logs in your logic apps should be enabled		25 of 26 logic apps		ASB		⚡ ⊙
Management ports of virtual machines should be protected with jus...		6 of 9 virtual machines		ASB, Azure CIS 1.1.0		⚡
Diagnostic logs in Key Vault should be enabled		2 of 4 key vaults		ASB, Azure CIS 1.1.0		⚡ ⊙
Web Application should only be accessible over HTTPS		1 of 3 web applicatio···		ASB, Azure CIS 1.1.0		⚡
⊓ Endpoint protection should be installed on your machines		2 of 8 virtual machines		ASB		⚡
Secure transfer to storage accounts should be enabled		1 of 8 storage accou···		ASB, Azure CIS 1.1.0		⚡ ⊖

Figure 4.31 – All recommendations – enforce action

6.   Select the **Storage account public access should be disallowed** recommendation:

Figure 4.32 – Fixing a recommendation

7.   On the recommendation page, identify three sections: **Description**, **Remediation steps**, and **Affected resources**.

The **Description** field explains what a recommendation is about, and why it is important to remediate recommended resources. The **Remediation steps** section displays the actual steps that you need to take to fix or remediate the affected resource or resources that are listed below the **Affected resources** section, under the **Unhealthy resources** tab.

8. Select one or more storage accounts to fix or to apply remediation steps.

9. Click **Fix**. A **Fixing resources** blade opens containing a remediation action description and a list of selected resources that will be remediated:

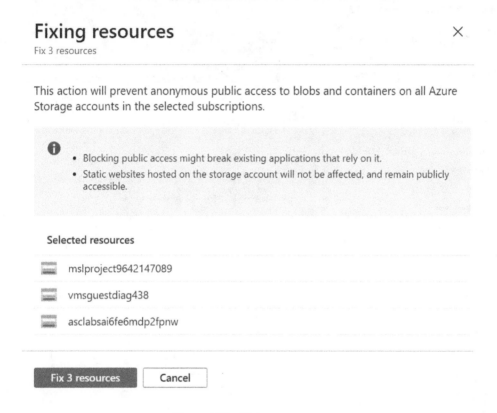

Figure 4.33 – Fixing resources

10. Click the blue button to fix the resources. The text on the button will be displayed as **Fix N resources**, where *N* represents the number of selected resources to fix.

How it works...

To fix unhealthy resources, Microsoft Defender for Cloud applies the changes to affected resources. To see what these changes are, read the **Remediation steps** section on the recommendation information blade.

Additionally, click **Quick fix logic** to see **Automatic remediation script content**, in JSON format:

Figure 4.34 – Automatic remediation script content

It can take several minutes after remediation completes to see the resources in the **Healthy resources** tab.

Managing a recommendation query in Azure Resource Graph Explorer

Microsoft Defender for Cloud uses Azure Policy to define security policy definitions, but for each security recommendation, you can see the appropriate Azure Resource Graph query.

In this recipe, you will learn how to manage a recommendation's **Azure Resource Graph** query.

Getting ready

Open a web browser and navigate to `https://portal.azure.com`.

How to do it...

To manage a recommendation's Azure Resource Graph query, take the following steps:

1. In the Azure portal, open **Microsoft Defender for Cloud**.
2. On the left menu, click **Recommendations**.
3. In the list of recommendations, search for and select the **Secure transfer to storage accounts should be enabled** recommendation.

4. At the top of the recommendation page, click on **Open a query**. The **Azure Resource Graph Explorer** blade opens. On the right side, under the **Query 1** tab, is a query that will return the recommendation result. Click **Run query**:

Figure 4.35 – Unhealthy resource status

5. The query executes and the result is listed under the **Results** tab.

6. In the results list, in the **status** column, resources that should be remediated have the **Unhealthy** status.

7. To save the query, select the **Save** or **Save as** button.

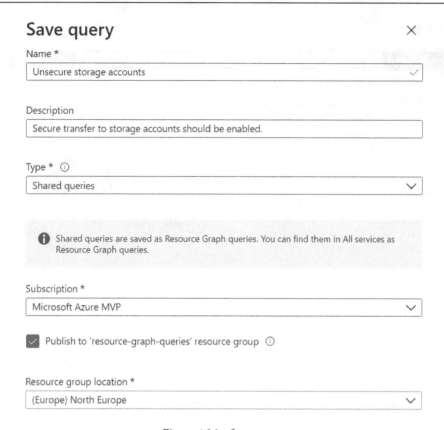

Figure 4.36 – Save query

8. Type a name and a description for the query.

9. From the **Type** menu, choose **Shared queries**. To make a query available only to you, choose **Private queries**.

10. Choose a subscription and a resource group location.

11. If the **Publish to 'resource-graph-queries' resource group** option is checked, a **resource-graph-queries** resource group is automatically created. If unchecked, you can create and name your own resource group. Leave the setting checked.

12. Click **Save**.

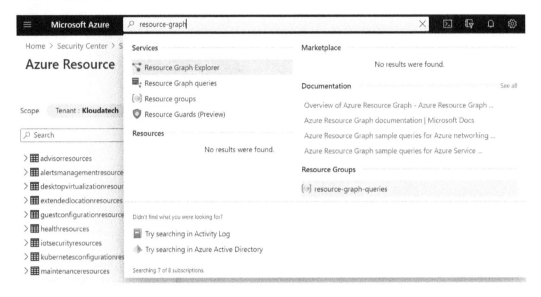

Figure 4.37 – Search for the resource-graph-query resource group

13. At the top of the Azure portal, in the search field, type `resource-graph` to search for a recently created resource group.

14. From the list of resource groups, select the **Resource-Graph-query** resource group:

Figure 4.38 – Saved Azure Resource Graph query

15. In the selected resource group, observe the **Global** location of the recently saved query. Select the query.

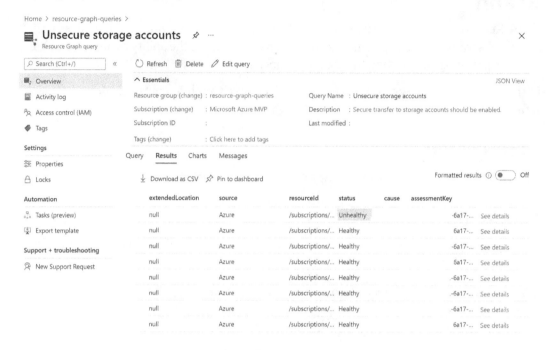

Figure 4.39 – Unhealthy resources query results

16. In the **Resource Graph query** blade, under the **Query** tab, is the query code. Click on the **Results** tab to run the query.

How it works...

A private query is only available to you and can be accessed only by you. A shared query is saved to a resource group, and you can control who has access to the shared query via role-based access control role assignments.

Getting a secure score using Azure Resource Graph

Azure Resource Graph is a service in Azure that is capable of querying resources at scale across subscriptions and exploring Azure resources, including the ability to apply complex filtering, grouping, and sorting by resource properties.

In this recipe, you will learn how to use Azure Resource Graph to query resources for the **Azure secure score** feature.

Getting ready

Open a web browser and navigate to `https://portal.azure.com`.

How to do it...

To filter, group, and sort security recommendations in Microsoft Defender for Cloud, take the following steps:

1. In the Azure portal, open **Resource Graph Explorer**.

2. On the left menu, under the **Scope** menu, choose a scope for the queries: **Tenant**, **Management Group**, or **Subscription**.

3. In the right pane, under the **Query 1** tab, type the following Kusto query:

    ```
    SecurityResources
    | where type == 'microsoft.security/securescores'
    | extend current = properties.score.current, max =
    todouble(properties.score.max)
    | project subscriptionId, current, max, percentage =
    ((current / max)*100)
    ```

4. Click **Run query**:

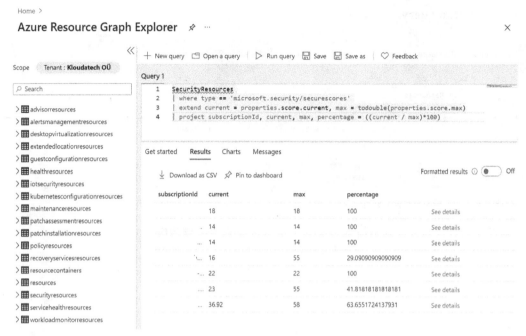

Figure 4.40 – Azure Resource Graph Explorer script results

5. The query returns the subscription ID, the maximum score for the subscription, as well as the current score value as a number and as a percentage.

    Click **New query**. A new **Query** tab will open.

6. In the **Query 2** tab, type the following Kusto query:

```
SecurityResources
| where type == 'microsoft.security/securescores/
securescorecontrols'
| extend SecureControl = properties.displayName,
unhealthy = properties.unhealthyResourceCount,
currentscore = properties.score.current, maxscore =
properties.score.max
| project SecureControl , unhealthy, currentscore,
maxscore
```

7.   Click **Run query**:

Figure 4.41 – Azure Resource Graph Explorer query results

8. The query returns the status of all security controls and the number of unhealthy resources for each control, as well as the current and maximum secure score.

How it works...

You can use **Kusto Query Language** (**KQL**) scripts to query Azure Resource Graph not just for the Azure secure score, but for numerous other objectives.

There's more...

To learn more about Azure Resource Graph, refer to the official documentation at `https://docs.microsoft.com/en-us/azure/governance/resource-graph/`.

5
Security Alerts

After you deploy resources in Azure on-premises and hybrid cloud environments, Microsoft Defender for Cloud can collect security-related data and display security alerts for the deployed resources. You need to enable Microsoft Defender for Cloud plans to enable advanced detections to trigger security alerts.

Once Microsoft Defender for Cloud plans are enabled, the Security Alerts blade in Microsoft Defender for Cloud will display relevant security alerts for deployed resources.

In this chapter, you will learn how to manage security alerts, respond automatically to alerts, use alert maps to view and manage alerts, create suppression rules, and remediate recommendations in Microsoft Defender for Cloud.

We will cover the following recipes in this chapter:

- Filtering, grouping, and exporting security alerts
- Responding to security alerts using automated responses
- Creating suppression rules
- Organizing security alerts and changing a security alert status

Technical requirements

To successfully complete the recipes in this chapter, the following is required:

- An Azure subscription

- A web browser, preferably Microsoft Edge

- Microsoft Defender for Cloud plans

- Resources in an Azure subscription, such as virtual machines, storage, SQL server, and Logic Apps. Microsoft Defender for Cloud will create resource recommendations based on available resources.

The code samples can be found at `https://github.com/PacktPublishing/Microsoft-Defender-for-Cloud-Cookbook`.

Filtering, grouping, and exporting security alerts

To get a more complete overview of the security posture of deployed resources, you can view security alerts in multiple ways. Also, you can export security alerts in CSV format for further analysis.

In this recipe, you will learn how to filter, group, and export security alerts, and change the display of the security alerts blade to view security alerts in different ways.

Getting ready

Open a web browser and navigate to `https://portal.azure.com`.

How to do it

To filter and group security alerts, and change how the security alerts blade displays security alerts in Microsoft Defender for Cloud, take the following steps:

1. In the Azure portal, open Microsoft Defender for Cloud.

2. To be able to see security alerts for resources, you must enable Microsoft Defender for Cloud plans. To enable Microsoft Defender for Cloud plans, view *Chapter 1, Getting Started with Microsoft Defender for Cloud*. If you have not enabled Microsoft Defender for Cloud plans, the Security alerts blade will look like this:

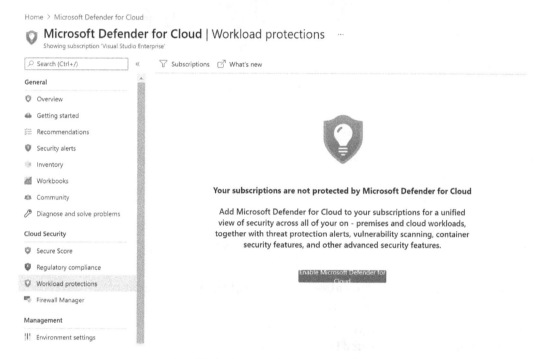

Figure 5.1 – Security alerts blade if Microsoft Defender for Cloud plans is not enabled

3. Once enabled, advanced detections in Microsoft Defender for Cloud will trigger security alerts and display them in the **Security alerts** blade. On the left menu, click **Security alerts**:

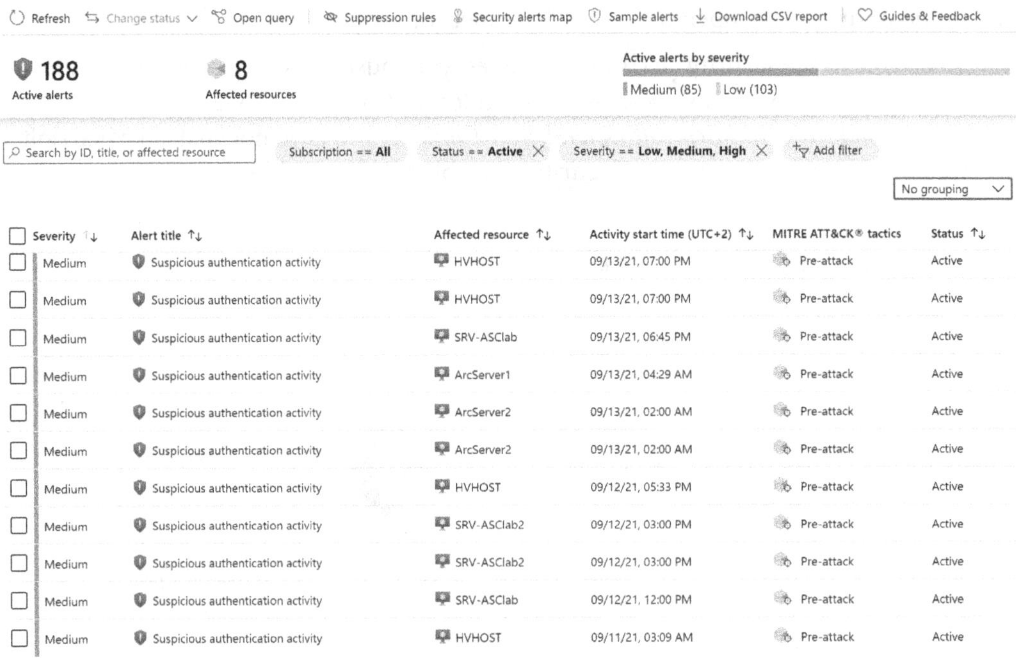

Figure 5.2 – Security alerts in Microsoft Defender for Cloud, after Microsoft Defender for Cloud plans is enabled

4. On the top menu, click **Download CSV report** to generate a report of the current security alerts in CSV file format.

5. After the download is finished, save the report. The security alerts report contains a number of fields: **exportedTimestamp**, **subscriptionId**, **subscriptionName**, **resourceGroup**, **resourceType**, **resourceNameresourceId**, **alertId**, **alertName**, **alertLocation**, **alertType**, **activityTimeUTC**, **alertDisplayName**, **alertSeverity**, **Description**, **remediationSteps**, **DetectedBy**, **azurePortalAlertLink**, and entities.

6. At the top of the page, in the search field, type to search alerts by ID, title, or affected resource. As you type, a filter is applied to the Security Alerts list. Type the name of a resource in one of your subscriptions covered by Microsoft Defender for Cloud to filter the Security Alerts list for the resource. Observe the filtered security list.

7. Delete any text in the search field.

8. On the right side of the search field, three filters are active by default: **Subscription**, **Status**, and **Severity**. Click on the **Subscription** filter:

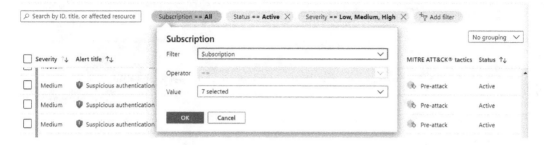

Figure 5.3 – Security alerts – Subscription filter

9. Click on **Value** to change the value of the **Filter** or click on the **Filter** drop-down menu to change the filter.

10. Click on **Add filter**, to add a new filter:

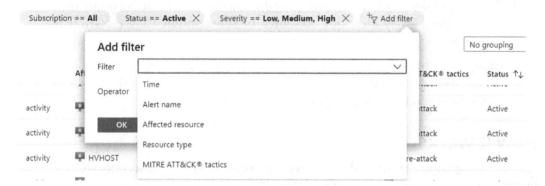

Figure 5.4 – Add filter

11. Click on the **Filter** drop-down menu to display the list of available filters: **Time, Alert name, Affected resource, Resource type**, and **MITRE ATT&CK® tactics**:

Figure 5.5 – Grouping security alerts

12. On the right side, click on the **Grouping** drop-down menu. You can group the Security Alerts list by **Alert title, Resource**, or **Subscription**. Click on **Group by alert title**.

13. Grouping security alerts allows you to find and examine security alerts in different views, raising your efficacy in triaging security alerts. Click on the **Group** drop-down menu and click **No grouping** to ungroup security alerts.

How it works

Filtering and grouping security alerts allows you to find and examine security alerts in multiple ways, such as filtering by subscription, status, time, alert name, affected resource, and resource type, or grouping by alert title, resource, or subscription.

Additionally, alert severity is categorized into four different levels: Informational, Low, Medium, and High, which provide help in prioritizing the approach to resolve alerts.

There's more

A reference guide to supported security alerts is available at `https://docs.microsoft.com/en-us/azure/defender-for-cloud/alerts-reference`.

Responding to security alerts using automated responses

In the process of prioritizing and responding to security alerts, you might want to respond to an alert in the form of an action, or a set of actions, preferably automated, that will remediate the security alert. Security alerts in Microsoft Defender for Cloud allow you to trigger an automated response to security alerts.

Getting ready

Open a web browser and navigate to `https://portal.azure.com`.

How to do it

To respond to security alerts using an automated response, complete the following steps:

1. In the Azure portal, open **Microsoft Defender for Cloud**.

2. In the left pane, click **Security Alerts**.

3. In the list of security alerts, click on an alert.

4. The details pane opens on the right. Click **View full details**:

Figure 5.6 – Security alert details

5. The security alert detailed information blade opens. On the left side, investigate **Severity**, **Status**, **Activity time**, **Alert description**, **Affected resources**, and the **kill chain intent** according to MITRE ATT&CK® framework tactics. The right side contains two tabs: **Alert details** and **Take action**. Investigate the information provided, as well as the **Related entities** information. Click **Next: Take Action**:

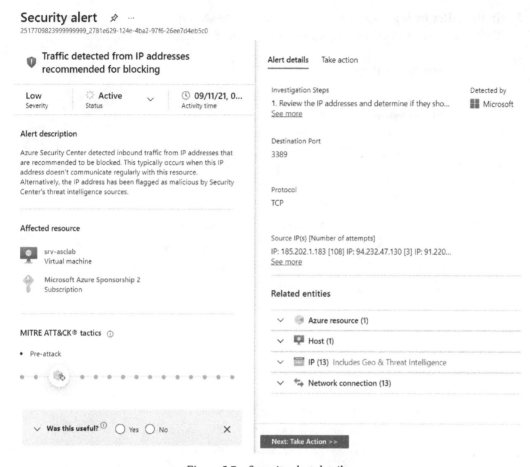

Figure 5.7 – Security alert details

6. The information on the **Take action** tab has several categories:

- **Mitigate the threat**: This is where you can click on **Enforce rule** to enforce a recommendation. See *Chapter 4, Secure Score and Recommendations.*

- **Prevent future attacks**: The list shows recommendations for a resource.

- **Trigger automated response**: You can trigger an automated response to a security alert.

- **Suppress similar alerts**: You can create suppression rules to suppress similar alerts. See recipes in this chapter to create suppression rules.

    Click on **Trigger Logic App**.

7. In the **Filter by logic app name** field, enter text to search and filter **Logic Apps**. In the **Select a subscription** drop-down menu, select an Azure subscription:

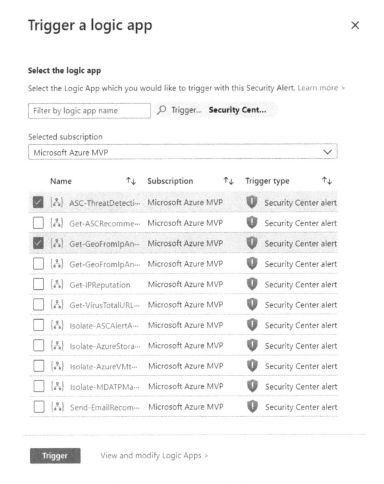

Figure 5.8 – Trigger a logic app

8. From the logic app list, select the logic apps you want to trigger or run.

9. Click **Trigger** to execute the selected logic apps.

How it works

If Microsoft Defender for Cloud identifies a security alert, its detailed information will help you determine the severity of the alert, as well as identifying and recommending steps to prevent future similar or identical attacks. The responses and remediation actions can involve multiple steps, and you can use one or more Logic Apps to automate a response to a security alert.

Creating suppression rules

The Security Alerts list can sometimes contain alerts that might not be relevant to you and that you don't want to be on the list. Typical reasons you would want to suppress alerts are alerts being triggered too often or if there are many false positives. In these cases, you might want to declutter the Security Alerts list, so you can focus on more relevant and important alerts.

In this recipe, you will learn how to create alert suppression rules, to temporarily suppress and remove alerts from the Security Alerts list.

Getting ready

Open a web browser and navigate to `https://portal.azure.com`.

How to do it

To create a suppression rule in Microsoft Defender for Cloud for a specific security alert, take the following steps:

1. In the Azure portal, open **Microsoft Defender for Cloud**.

2. On the left menu, click **Security alerts**.

3. Select a security alert you want to create a suppression rule for. The details pane opens on the right side. Click **Take action**.

4. On the right side of the security alert blade, click on the **Suppress similar alerts** category to expand it. Click **Create Suppression Rule**:

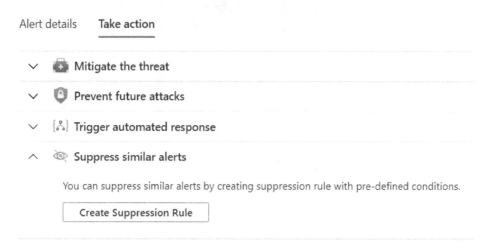

Figure 5.9 – Take action tab – Create Suppression Rule

5. A **New suppression rule** window opens on the right. It contains three sections: **Rule conditions**, **Rule details**, and **Rule expiration**.

    In the **Rule conditions** section, select one or more subscriptions:

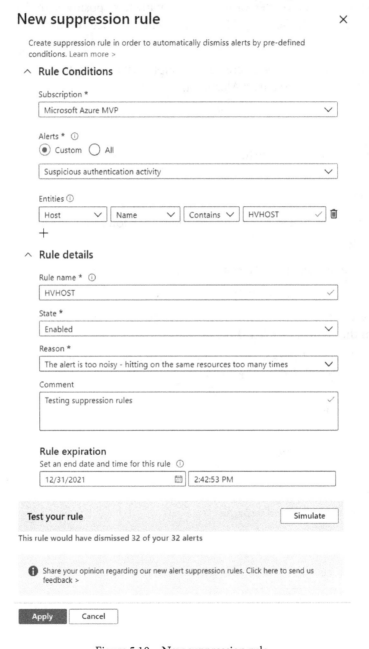

Figure 5.10 – New suppression rule

The **Alerts** drop-down menu contains alerts that have been triggered at least once on the selected subscriptions. In this case, it will contain only the alert that corresponds to the category of alert you selected in *step 3*. Instead, you can click on the **All** radio button to apply a suppression rule to all the alerts in the selected subscriptions.

The **Entities** fields contain options to refine suppression rules to specific entities. These drop-down menus will differ based on a selected suppression alert type. For example, if the **Suspicious authentication activity** alert is selected, you can select **Host** or **Azure Resource** for an entity type, and different values based on the entity type, such as **Name, Azure id, DNS domain**, and others.

Under the **Rule details** section, type a name in the **Rule name** field.

6. Make a selection from the **State** dropdown for the state of the rule: by default it is **Enabled**, but **Disabled** and **Expired** are available as well.

7. From the **Reason** dropdown, select the reason you are creating a suppression rule:

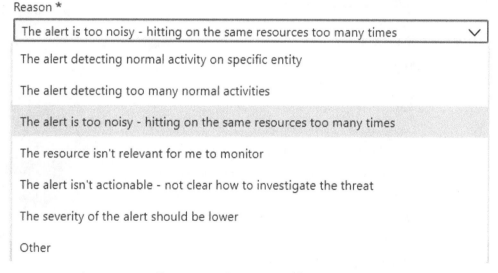

Figure 5.11 – Suppression rule reason

8. In the **Comment** field, explain why you are creating a suppression rule.

9. Under the **Rule expiration** section, define an end date and time for the rule. Suppression rules are not supposed to be permanently enabled, for security reasons. Therefore, the rule expiration date is limited to a maximum of 6 months from the creation date.

10. The **Test your rule** section contains an option to test your rule and see what the impact will be when enabled. Click **Simulate**. As a result, you will see information on how many alerts the rule would have dismissed if it had been active.

11. Click **Apply** to apply the changes and activate the rule.

12. Close all blades and return to the **Security alerts** blade. On the right side, on the top menu, click **Suppression rules**. The **Suppression Rules** blade opens. Here, you can see all the suppression rules you have defined, edit existing suppression rules, and create new suppression rules.

    To create a suppression rule, click **+ Create new suppression rule**:

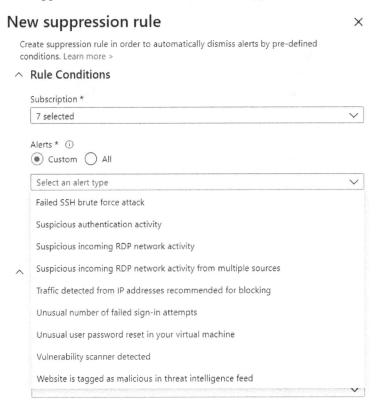

Figure 5.12 – Create a new suppression rule dialog

13. The procedure to create a suppression rule is the same as described from *step 5* onwards, except for the possibility to choose any alert from a list of the available alerts.

How it works

To dismiss alerts that are not important or interesting, you will create suppression rules. To create a suppression rule, you define the criteria for which alerts will be automatically dismissed, and these could be based on any alert type that has been triggered at least once in target subscriptions. After that, you can define a rule more granularly, defining to which entities it will apply.

There's more

It is important that you know the implications of creating suppression rules.

> **Important Note**
> When you create suppression rules and prevent alerts from being displayed in the Security Alerts list, you diminish the efficacy of Microsoft Defender for Cloud. This might decrease the overall protection and lower the security posture of your infrastructure. You should monitor suppression rules and suppressed alerts and revise them as necessary.

Organizing security alerts and changing a security alert status

Security alerts are generated continuously based on advanced analytics and threat intelligence, and when a security alert is raised, you should respond to it and resolve it as soon as possible. In this recipe, you will learn how to respond to a security alert and change its alert status.

Getting ready

Open a web browser and navigate to `https://portal.azure.com`.

How to do it

To respond to a security alert and change its alert status, take the following steps:

1. In the Azure portal, open **Microsoft Defender for Cloud**.
2. On the left menu, select **Security alerts**.

3. Select a security alert you want to investigate and remediate. A details pane opens on the right. At the top of the details pane, click on the **Status** drop-down menu, showing the alert status as **Active**:

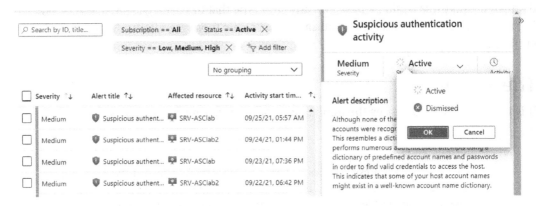

Figure 5.13 – Security alert status

4. If a security alert is resolved, you should change its status, which will remove it from the list of active security alerts. Click on **Dismissed** and select **OK**. Alternatively, you can click on **View full details** and change the status of an alert from there as well, from the same menu:

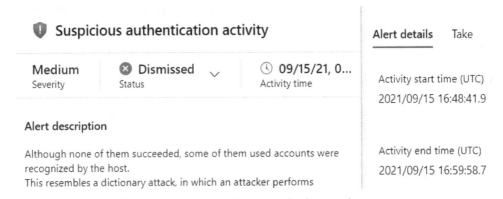

Figure 5.14 – An alert displaying the dismissed status

5. After you dismiss an alert, its status will show as **Dismissed**.

6. Close all blades to return to the **Security alerts** blade.

7. To change the security alert status and dismiss one or more alerts at once, select the checkboxes of the alerts you want to dismiss.

8. On the top menu, click on **Change status**, and select **Dismissed**:

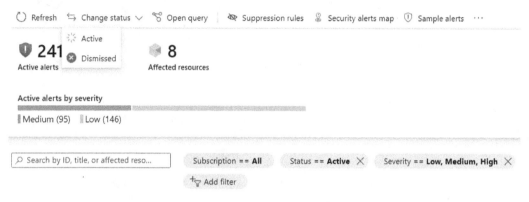

Figure 5.15 – Change an alert status from the top menu

9. To show the list of dismissed alerts, change the **Status** filter from **Active** to **Dismissed**:

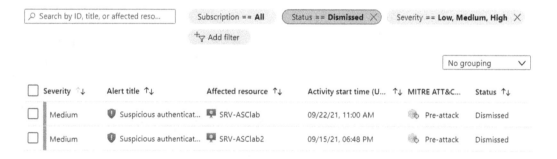

Figure 5.16 – List of dismissed alerts

10. Occasionally, you may want to change the status of a dismissed alert back to **Active**. In that case, while displaying the list of dismissed alerts, change the alert status in the same way as described in this recipe.

How it works

When responding to security alerts, you will want to remove resolved alerts from the list of active alerts to reduce the amount of information, and changing a security alert's status accomplishes the goal.

6
Regulatory Compliance and Security Policy

In this chapter, you will learn how to manage security policies, manage compliance recommendations, add regulatory and compliance standards, manage compliance controls, and improve regulatory compliance in Microsoft Defender for Cloud.

Microsoft continuously updates regulatory compliance standards in Microsoft Defender for Cloud, displayed in the regulatory compliance dashboard. Using compliance assessment results presented in the dashboard, you can see the result of comparing the configuration of your resources with regulations, benchmarks, and industry standards. This way, you can quickly and efficiently remediate the results and work toward meeting particular compliance requirements.

We will cover the following recipes in this chapter:

- Managing Microsoft Defender for Cloud's default security policy
- Adding a custom security initiative and policy
- Adding a regulatory compliance standard

- Improving regulatory compliance, exempting, and denying a compliance control
- Accessing and downloading compliance reports

Technical requirements

To successfully complete recipes in this chapter, the following is required:

- An Azure subscription.
- A web browser, preferably Microsoft Edge.
- Microsoft Defender for Cloud plans.
- Resources in Azure subscription, such as virtual machines, storage, SQL Server, and Logic Apps. Microsoft Defender for Cloud will create resource recommendations based on available resources.

The code samples can be found at `https://github.com/PacktPublishing/Microsoft-Defender-for-Cloud-Cookbook`.

Managing Microsoft Defender for Cloud's default security policy

Microsoft Defender for Cloud generates security recommendations based on security policies. A security policy can be defined either on a management group, subscription, or a resource group.

In this recipe, you will learn to manage Microsoft Defender for Cloud's default security policy on a management group and a subscription.

Getting ready

Open a web browser and navigate to `https://portal.azure.com`.

How to do it...

To manage the Microsoft Defender for Cloud's default security policy, complete the following steps:

1. In the Azure portal, open **Microsoft Defender for Cloud**.
2. In the left menu, click **Environment settings**.

3. Click on an arrow next to a management group to display a list of associated subscriptions. If you have created management groups and assigned subscriptions to management groups, the **Environment settings** page will display a list of management groups, with associated subscriptions, as shown in this example:

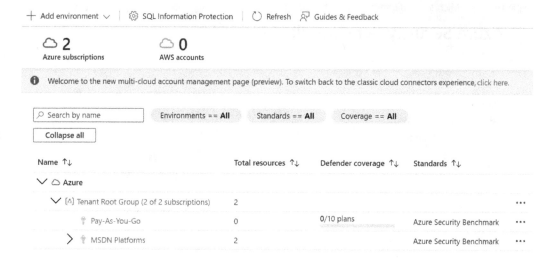

Figure 6.1 – Environment settings

4. Click on a subscription.

5. The **Defender Plans** blade opens. In the left menu, click on **Security Policy**.

6. On the **Security Policy** blade, examine the **Default initiative** section and the list of default initiatives enabled on the selected subscription:

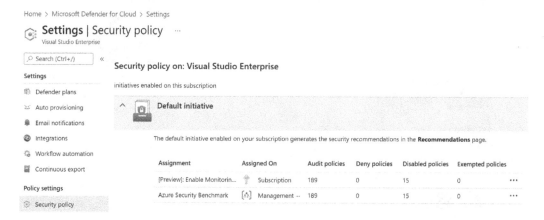

Figure 6.2 – Security policy

7. On the right side of an initiative, click on the ellipsis to display a menu. The menu gives you the possibility to delete an initiative assignment, edit an initiative assignment, or create an exemption.

    Select **Edit assignment**:

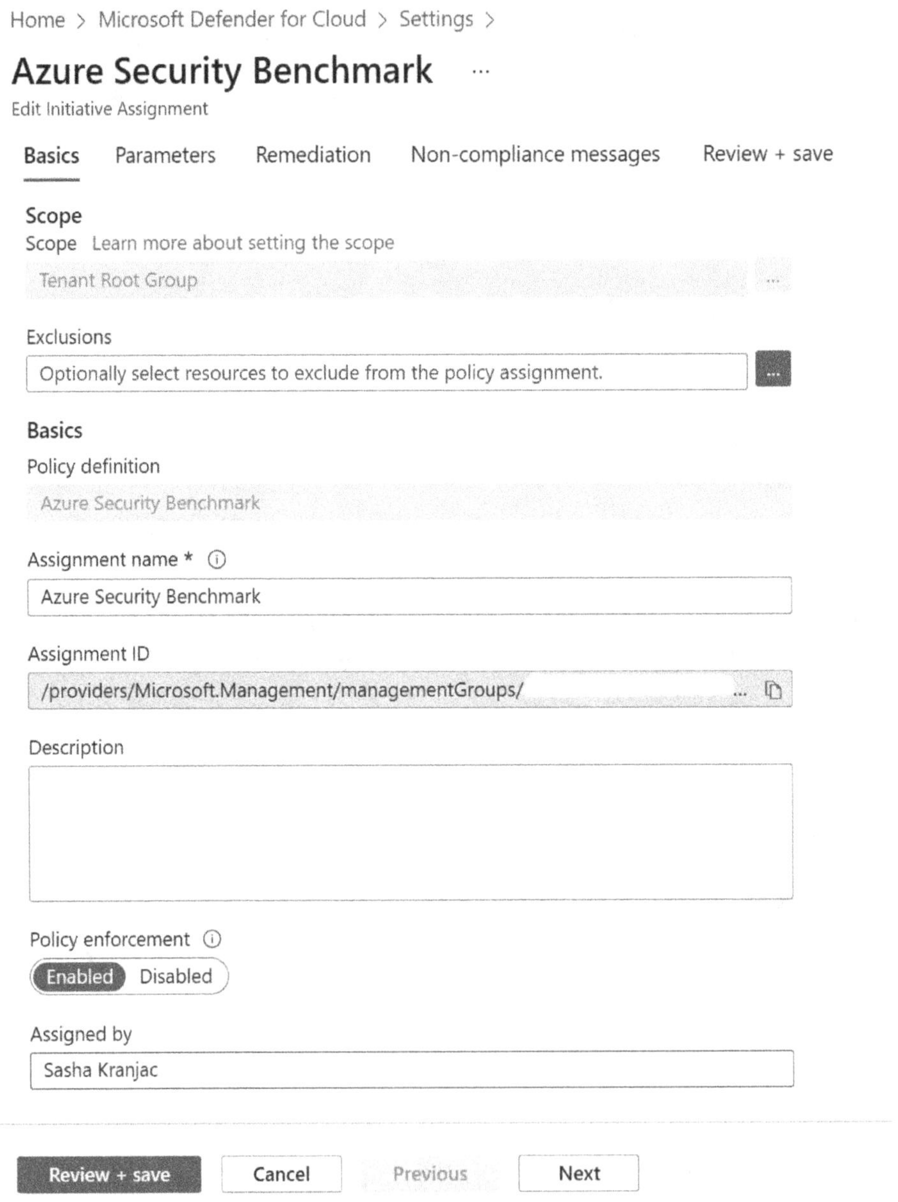

    Home > Microsoft Defender for Cloud > Settings >

    # Azure Security Benchmark   ...
    Edit Initiative Assignment

    Basics   Parameters   Remediation   Non-compliance messages   Review + save

    Scope
    Scope Learn more about setting the scope

    Tenant Root Group   ...

    Exclusions

    Optionally select resources to exclude from the policy assignment.   ...

    Basics
    Policy definition

    Azure Security Benchmark

    Assignment name * ⓘ

    Azure Security Benchmark

    Assignment ID

    /providers/Microsoft.Management/managementGroups/   ... 🗗

    Description

    Policy enforcement ⓘ
    (Enabled Disabled)

    Assigned by

    Sasha Kranjac

    [Review + save]   [Cancel]   [Previous]   [Next]

    Figure 6.3 – Edit Initiative Assignment

8. On the **Edit Initiative Assignment** blade, you can make several changes to an initiative. On the **Basics** page, in **Exclusions**, you can optionally exclude the following resources from the policy assignment: management groups, subscriptions, resource groups, and resources.

    Click the ellipsis on the right side of the **Exclusions** text field.

9. On the **Exclusions** blade, select a subscription and a resource group. Click the **Add to selected scope** button. Click **Save** to save the selected exclusions.

10. The **Policy enforcement** switch allows you to disable policy enforcement. When policy enforcement is disabled, the policy effect is not enforced, but compliance assessment results are still available. Click **Next**.

11. On the **Parameters** page, unselect **Only show parameters that need input or review** to display policy parameters that you can edit and change individually. Click **Next**.

12. On the **Remediation** page, you can select either **System assigned managed identity** or **User assigned managed identity**. You will need a managed identity with policies containing **deployIfNotExists** and **modify** effect types to be able to deploy resources and edit tags on existing resources. Click **Next**.

13. The **Non-compliance messages** page enables you to define a default non-compliance message. To define a policy-specific non-compliance message, click on an ellipsis right of a policy definition name and click **Edit message**:

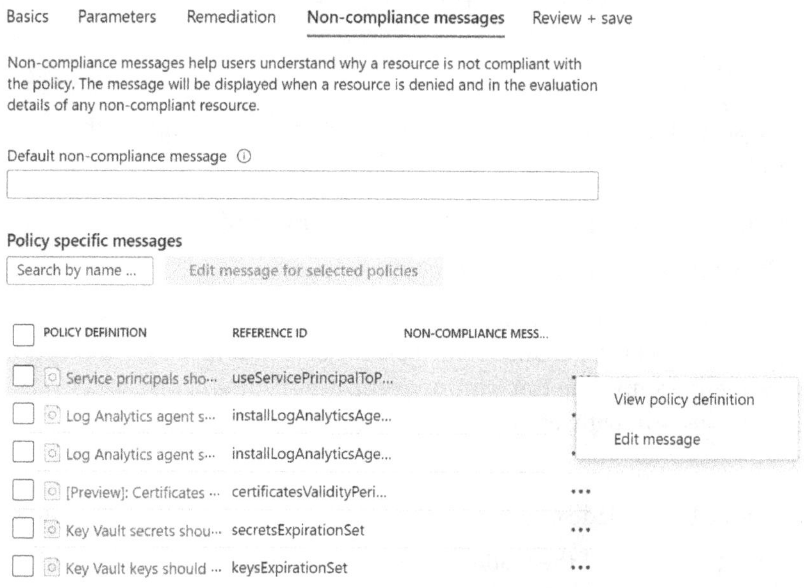

Figure 6.4 – Non-compliance messages

14. Type a non-compliance message and click **Save**.

15. Click **Next**. On the **Review + Save** page, review the changes you made and click **Save**.

How it works...

When Defender for Cloud is enabled on a subscription, the Azure Security Benchmark policy initiative is automatically assigned to the subscription. Based on the policy parameters, the security policy initiative takes effect on a subscription. Furthermore, recommendations are generated and actions applied.

The policies' effect can be **Append**, **Audit**, **AuditIfNotExists**, **Deny**, **DeployIfNotExists**, and **Disabled**.

Adding a custom security initiative and policy

Microsoft Defender for Cloud supports adding your own custom initiatives.

In this recipe, you will learn to add a custom security initiative and regulatory compliance policy.

Getting ready

Open a web browser and navigate to `https://portal.azure.com`.

How to do it...

To add a custom security initiative and policy to a subscription, complete the following steps:

1. In the Azure portal, open **Microsoft Defender for Cloud**.

2. In the left-hand menu, click **Environment settings**.

3. Click on an arrow next to a management group to display a list of associated subscriptions. If you have created management groups and assigned subscriptions to management groups, the **Environment settings** page will display a list of management groups, with associated subscriptions.

4. Click on a subscription.

5. The **Defender Plans** blade opens. In the left menu, click on **Security Policy**.

6. Scroll down to the end of the blade until the **Your custom** initiatives section is visible.

7. Click **Add a custom initiative**.

8. At the **Add custom initiatives** blade, click **+ Create new** to create a new custom policy initiative.

9. On the **Basics** page, click on an ellipsis right next to **Initiative location** and choose a location where the initiative will be stored. The initiative assignment is only available to resources at or below this location in the hierarchy.

10. Enter a name and a description of the initiative definition.

11. Use an existing category or create a new category for your initiative. Optionally, you can define an initiative version as well. Click **Next**.

12. On the **Policies** page, click **Next** to skip this page and navigate to **Controls**.

13. Click the **Create Control** button and create a control; define a name, control domain, and add a description. Once you have created a control, you will be able to return to the **Policies** page and add a control to a policy. Save the control and click **Previous** to return to **Policies**.

14. Click **Add policy definition**.

15. Choose one or more policies or controls and click **Add**.

16. Select one or more policies and click the **Add selected policies to a control** button:

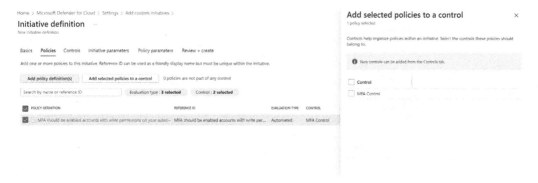

Figure 6.5 – Initiative definition – Policies

17. On **Add selected policies to a control**, select which controls the selected policies should belong to and click **Save**.

18. Click **Next** to navigate to the **Controls** page. Click **Next** to navigate to the subsequent pages – **Initiative parameters** and **Policy parameters**.

    Optionally, add initiative and (or) policy parameters. For example, **allowedLocations** is a parameter that enables a policy to limit the location of resources to a specified location only.

19. Click **Review + create** to complete creating the initiative definition.

20. On the right side of the **Add custom initiatives** page, click **Add** to add a policy to a subscription:

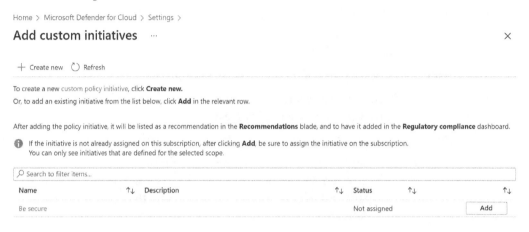

Figure 6.6 – Add custom initiatives

21. Review the policy settings on each page (**Basics, Parameters, Remediation,** and **Non-compliance messages**) and click **Review + create**:

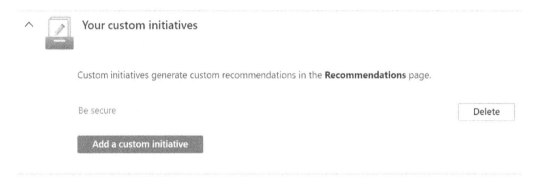

Figure 6.7 – The custom initiatives list

How it works...

Once you add your own custom initiative, you will begin to receive recommendations based on the custom security initiative, along with already defined and associated security initiatives.

If there is a conflict between two or more security initiatives, you will see a warning message and a link to expand the list of policy assignments in conflict:

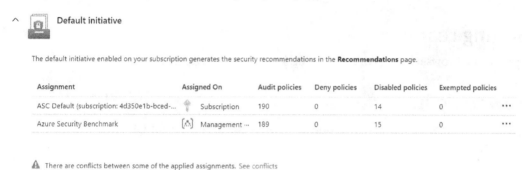

Assignment		Assigned On	Audit policies	Deny policies	Disabled policies	Exempted policies	
ASC Default (subscription: 4d350e1b-bced-...		Subscription	190	0	14	0	...
Azure Security Benchmark		Management ...	189	0	15	0	...

⚠ There are conflicts between some of the applied assignments. See conflicts

Figure 6.8 – Security initiatives enabled on a subscription

Click on **See conflicts** to expand and list the details of recommendation conflicts:

Security policy on: Courses

initiatives enabled on this subscription

∧ 🔒 **Default initiative**

The default initiative enabled on your subscription generates the security recommendations in the **Recommendations** page.

Assignment		Assigned On	Audit policies	Deny policies	Disabled policies	Exempted policies	
ASC Default (subscription: 4d350e1b-bced-...		Subscription	190	0	14	0	...
Azure Security Benchmark		Management ...	189	0	15	0	...

Recommendation conflict	Effective state	Details
Audit unrestricted network access to storage accounts	Audit	Disabled on Azure Security Benchmark

Figure 6.9 – Security initiatives – recommendation conflicts

Adding a regulatory compliance standard

Microsoft Defender for Cloud assigns the Azure Security Benchmark regulatory compliance standard as default with every subscription. You can add a regulatory standard initiative to Azure subscriptions from the growing list of supported regulatory standards.

In this recipe, you will learn to add a regulatory compliance standard to an Azure subscription.

Getting ready

Open a web browser and navigate to `https://portal.azure.com`.

How to do it...

To add a regulatory compliance standard to Microsoft Defender for Cloud, complete the following steps:

1. In the Azure portal, open **Microsoft Defender for Cloud**.
2. On the left menu, click **Environment settings**.
3. On the **Environment settings** blade on the right, select a subscription to open the **Settings** blade.
4. On the left menu, click on **Security policy**.
5. **Security policy** contains three sections – **Default initiative**, **Industry & regulatory standards**, and **Your custom initiatives**:

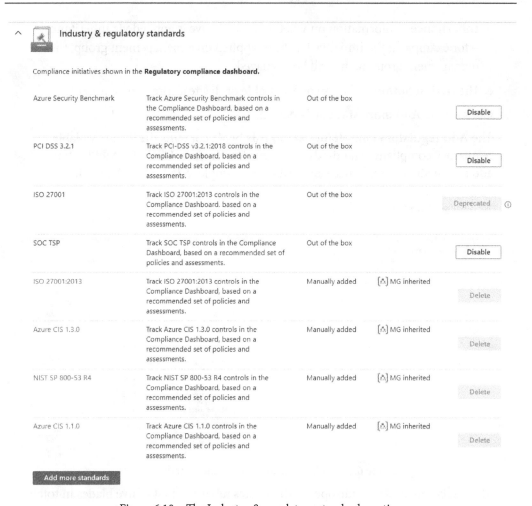

Figure 6.10 – The Industry & regulatory standards section

6. The **Industry & regulatory standards** section shows the active compliance initiatives that are shown in the Regulatory compliance dashboard. The list contains the following information about compliance initiatives:

    a. **Name** – identifies a compliance initiative and provides a link to the initiative definition

    b. **Description** – a short initiative and standard description

    c. **Type** – the type of the policy, either **Out of the box** or **Manually added**

d. **Inheritance** – information on whether an initiative is applied via inheritance – for example, if the initiative has been applied on a management group, the management group name will be displayed

e. **The Action button** – allows you to disable or delete a compliance initiative

7. Click on the **Add more standards** button.

8. The **Add regulatory compliance standards** blade contains the list of available regulatory compliance standards and their descriptions. Click the **Add** button to add a regulatory compliance standard – for example, **NIST SP 800-53 R5**:

Add regulatory compliance standards   ...   ×

Click **Add** on the standards that you want to add to the regulatory compliance dashboard and then assign it to the subscription. After completing the assignment , the custom policies will be available in the **Regulatory compliance** dashboard.

Search to filter items...			
Name ↑↓	Description ↑↓	↑↓	↑↓
NIST SP 800 171 R2	Track NIST SP 800 171 R2 controls in the Compliance Dashboard, based on a recommended set...		Add
UKO and UK NHS	Track UK OFFICIAL and UK NHS controls in the Compliance Dashboard, based on a recommend...		Add
Canada Federal PBMM	Track Canada Federal PBMM controls in the Compliance Dashboard, based on a recommended ...		Add
HIPAA HITRUST	Track HIPAA/HITRUST controls in the Compliance Dashboard, based on a recommended set of ...		Add
SWIFT CSP CSCF v2020	Track SWIFT CSP CSCF v2020 controls in the Compliance Dashboard, based on a recommended...		Add
New Zealand ISM Restricted	Track New Zealand ISM Restricted controls in the Compliance Dashboard, based on a recomme...		Add
CMMC Level 3	Track CMMC Level 3 controls in the Compliance Dashboard, based on a recommended set of p...		Add
NIST SP 800-53 R5	Track NIST SP 800-53 R5 controls in the Compliance Dashboard, based on a recommended set ...		Add
FedRAMP H	Track FedRAMP H controls in the Compliance Dashboard, based on a recommended set of poli...		Add
FedRAMP M	Track FedRAMP H controls in the Compliance Dashboard, based on a recommended set of poli...		Add

Figure 6.11 – Add regulatory compliance standards

9. The **Assign initiative** blade opens. The **Basics** tab is the first of five blades in total; **Scope** reflects the current working subscription, while you can add any exclusions to the initiative and a description. Click **Next**.

10. On the **Parameters** tab, you can search for and edit any initiative parameter. Click **Next**.

11. On the **Remediation** tab, you can configure remediation tasks for existing and newly created resources. To enable the initiative to apply **deployIfNotExist** and **modify** types to resources in a subscription, you need a managed identity, and here, you can choose between **System assigned managed identity** and **User assigned managed identity**. Click **Next**:

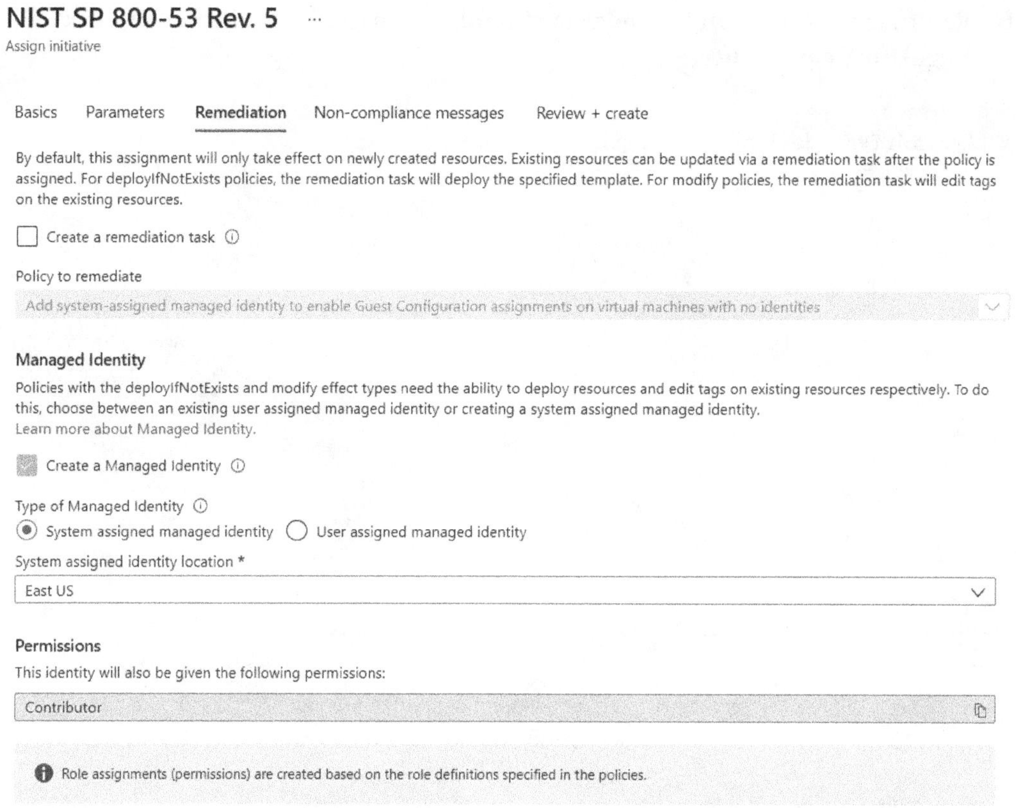

NIST SP 800-53 Rev. 5   ...
Assign initiative

Basics   Parameters   **Remediation**   Non-compliance messages   Review + create

By default, this assignment will only take effect on newly created resources. Existing resources can be updated via a remediation task after the policy is assigned. For deployIfNotExists policies, the remediation task will deploy the specified template. For modify policies, the remediation task will edit tags on the existing resources.

☐ Create a remediation task ⓘ

Policy to remediate

Add system-assigned managed identity to enable Guest Configuration assignments on virtual machines with no identities   ⌄

Managed Identity

Policies with the deployIfNotExists and modify effect types need the ability to deploy resources and edit tags on existing resources respectively. To do this, choose between an existing user assigned managed identity or creating a system assigned managed identity.
Learn more about Managed Identity.

☑ Create a Managed Identity ⓘ

Type of Managed Identity ⓘ
◉ System assigned managed identity   ○ User assigned managed identity
System assigned identity location *

East US   ⌄

Permissions

This identity will also be given the following permissions:

Contributor   ⎘

ⓘ Role assignments (permissions) are created based on the role definitions specified in the policies.

Figure 6.12 – Assign initiative – the Remediation tab

12. On the **Non-compliance messages** tab, you can review policy-specific messages that are displayed in case you have any non-compliant resources or when a resource is denied. Click **Next**.

13. On the **Review + create** page, review the initiative information and click **Create** to assign the initiative.

14. Return to the **Microsoft Defender for Cloud** main page. On the left menu, click **Regulatory compliance**:

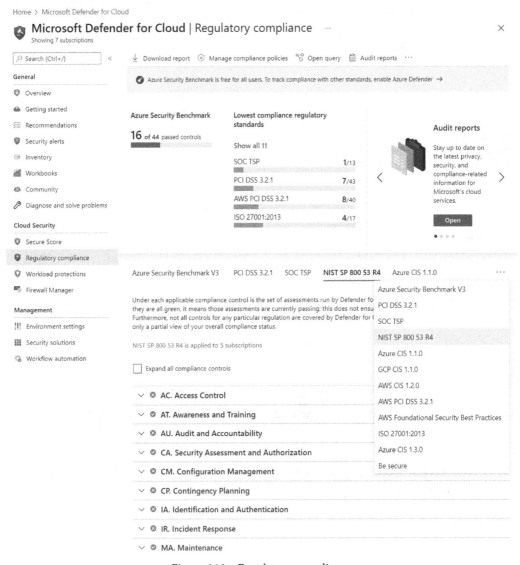

Figure 6.13 – Regulatory compliance

15. The top part of the **Regulatory compliance** page shows the **Azure Security Benchmark** score, in addition to other compliance regulatory standards' scores, sorted from lowest to highest score.

    The lower part shows a set of assessments associated with the chosen compliance control. If a recently added compliance control is not displayed on the page, on the right edge of the page, click on the ellipsis to show additional compliance controls and select the desired one.

16. The list shows a set of assessments associated with the selected control run by Defender for Cloud, organized in categories. Click on an arrow next to a category to expand and see the associated assessments:

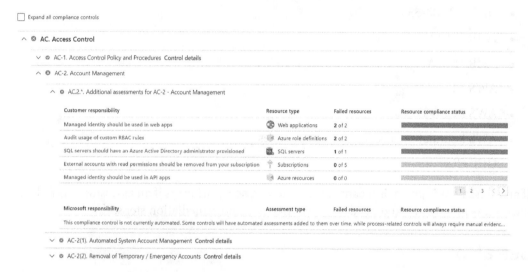

Figure 6.14 – The regulatory compliance control details

How it works...

Presently, the following regulatory compliance standards are available:

- PCI-DSS v3.2.1:2018

- SOC TSP

- NIST SP 800-53 R4

- NIST SP 800 171 R2

- UK OFFICIAL and UK NHS

- Canada Federal PBMM

- Azure CIS

- HIPAA/HITRUST

- SWIFT CSP CSCF v2020

- ISO 27001:2013

- New Zealand ISM Restricted

- CMMC Level 3

- NIST SP 800-53 R5

- FedRAMP H

- FedRAMP M

- AWS CIS

- AWS PCI DSS

- AWS Foundational Security Best Practices

- GCP CIS

If an assessment is green, it means it has a passing score, whereas if an assessment is red, it means it is not passing and needs further attention and remediation steps.

See also

Microsoft provides information about regulatory and compliance standards in Azure on various pages. Detailed information, downloadable information, as well as comprehensive information about regulatory compliance standards, can be accessed online, such as a standards overview, applicability, and services in scope. For your convenience, in the following list, we provide a few links to get started with compliance, governance, and regulatory standards:

- Azure compliance documentation: `https://docs.microsoft.com/en-us/azure/compliance/`

- Azure Global Compliance Map: `https://azure.microsoft.com/en-us/resources/azure-global-compliance-map/`

- Introduction to regulatory compliance: `https://docs.microsoft.com/en-us/azure/cloud-adoption-framework/govern/policy-compliance/regulatory-compliance`

- Microsoft Trust Center: `https://www.microsoft.com/en-us/trust-center`

- Microsoft Service Trust Portal: `https://servicetrust.microsoft.com/`

Improving regulatory compliance, exempting, and denying a compliance control

Microsoft Defender for Cloud continuously monitors and assesses your environment corresponding to regulatory compliance controls applied to subscriptions.

In this recipe, you will learn how to improve your regulatory compliance.

Getting ready

Open a web browser and navigate to `https://portal.azure.com`.

How to do it...

To improve your regulatory compliance status, complete the following steps:

1. In the Azure portal, open **Microsoft Defender for Cloud**.
2. On the left menu, click **Regulatory compliance**.
3. On the tabs showing names of regulatory compliance standards, click on **Azure Security Benchmark**. We are using this regulatory compliance policy as an example – you are welcome to use any regulatory compliance you want.

4. To expand a compliance control category, click on an arrow next to the desired category. There are two squares, containing **MS** and **C** letters. **MS** means a category contains controls that are under Microsoft's responsibility, while **C** means a category contains controls that are a part of the customer's responsibility:

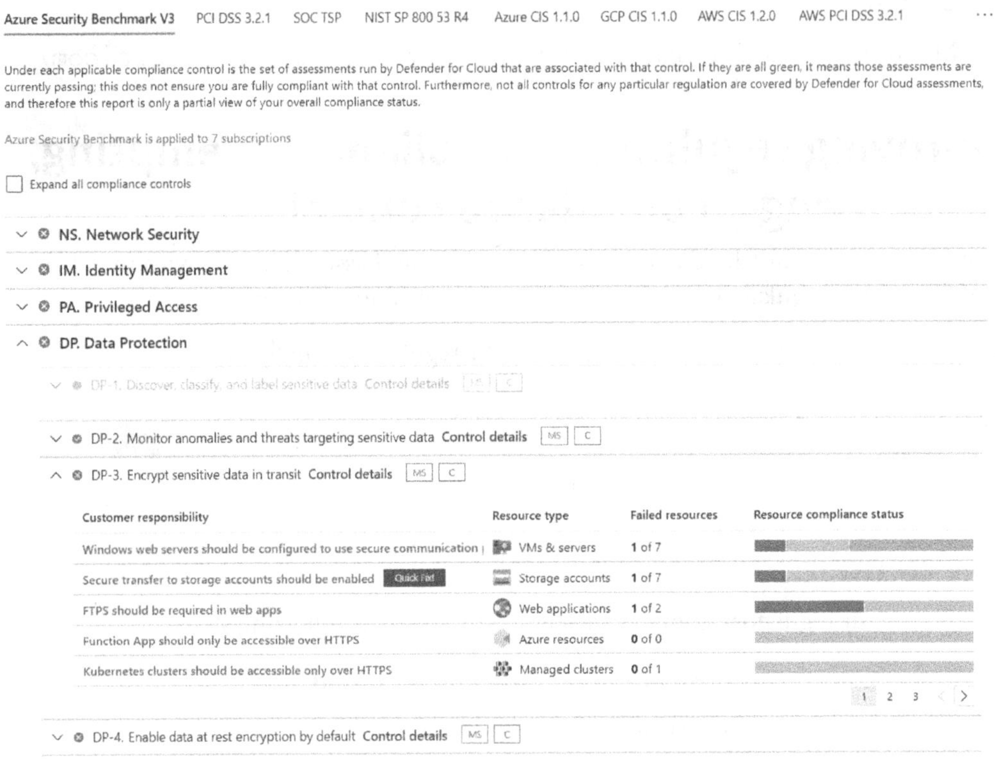

Azure Security Benchmark V3   PCI DSS 3.2.1   SOC TSP   NIST SP 800 53 R4   Azure CIS 1.1.0   GCP CIS 1.1.0   AWS CIS 1.2.0   AWS PCI DSS 3.2.1   ⋯

Under each applicable compliance control is the set of assessments run by Defender for Cloud that are associated with that control. If they are all green, it means those assessments are currently passing; this does not ensure you are fully compliant with that control. Furthermore, not all controls for any particular regulation are covered by Defender for Cloud assessments, and therefore this report is only a partial view of your overall compliance status.

Azure Security Benchmark is applied to 7 subscriptions

☐ Expand all compliance controls

∨ ⊗ NS. Network Security

∨ ⊗ IM. Identity Management

∨ ⊗ PA. Privileged Access

∧ ⊗ DP. Data Protection

  ∨ ⊚ DP-1. Discover, classify, and label sensitive data Control details MS C

  ∨ ⊚ DP-2. Monitor anomalies and threats targeting sensitive data Control details MS C

  ∧ ⊗ DP-3. Encrypt sensitive data in transit Control details MS C

Customer responsibility	Resource type	Failed resources	Resource compliance status
Windows web servers should be configured to use secure communication	VMs & servers	**1** of 7	
Secure transfer to storage accounts should be enabled Quick Fix	Storage accounts	**1** of 7	
FTPS should be required in web apps	Web applications	**1** of 2	
Function App should only be accessible over HTTPS	Azure resources	**0** of 0	
Kubernetes clusters should be accessible only over HTTPS	Managed clusters	**0** of 1	

1 2 3 ＜ ＞

  ∨ ⊗ DP-4. Enable data at rest encryption by default Control details MS C

Figure 6.15 – An expanded compliance control

5. In this example, the compliance status of the **Secure transfer to storage accounts should be enabled** control is not adequate, and given its compliance status, we might want to improve its compliance. Under **Customer responsibility**, click on the **Secure transfer to storage accounts should be enabled** control link.

6.   The selected control's page opens:

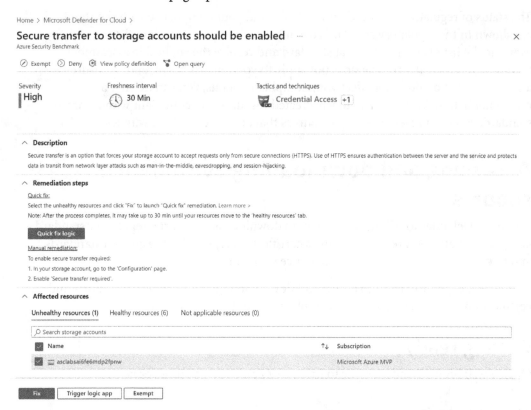

Figure 6.16 – Security control details

7.   In the **Affected resources** section, select the resources you want to remediate. After you have selected a resource, three buttons at the bottom of the page become active.

You can click **Fix** to remediate selected resources, **Trigger logic app** to execute a logic app with this recommendation, or **Exempt** to exempt a recommendation from any scope so that it does not affect the secure score.

8.   At the top of the page, you can click the **Deny** button to set the scope to block the creation of resources that do not comply with the recommendation.

How it works...

The status of regulatory standards assessments and your compliance with the standards are shown in the regulatory standards dashboard. To improve the compliance, examine the list of controls for that standard and resolve the applicable assessment recommendations. If a control does not apply to your environment, and to ensure a recommendation does not affect your compliance status, you can exempt a recommendation. To make sure newly created resources are compliant with a current standard, you can prevent creating resources that are not compliant with a standard.

Accessing and downloading compliance reports

Microsoft Defender for Cloud allows you to download and export a regulatory standard compliance status, Azure and Dynamics certification reports for the applied standards, and to view and track regulatory compliance over time.

In this recipe, you will learn to access and download regulatory standards reports and certification reports, and check the compliance over time dashboard.

Getting ready

Open a web browser and navigate to `https://portal.azure.com`.

How to do it...

To download reports and check regulatory compliance status over time, complete the following steps:

1. In the Azure portal, open **Microsoft Defender for Cloud**.

2. On the left menu, click **Regulatory compliance**.

3. To download a regulatory compliance status report, in the top menu, click the **Download report** button. The **Download report** blade opens on the right.

4. From the **Report standard** menu, select a regulatory standard. The list of regulatory standards contains active standards on your subscriptions.

5. From the **Format** menu, select the **CSV** or **PDF** report format:

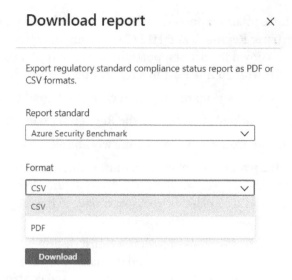

Figure 6.17 – Download report

6. Click **Download** to download and save a report.

7. Close the **Download report** blade.

8. In the top menu, click the **Audit reports** button. The **Audit reports** blade opens:

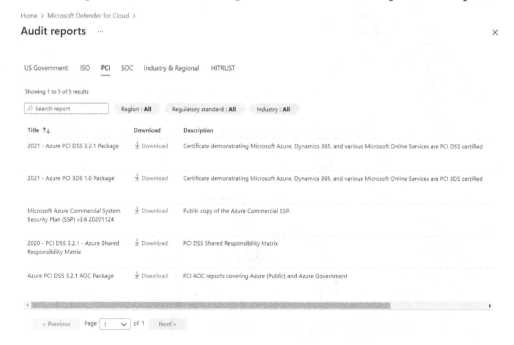

Figure 6.18 – Audit reports

9. Each tab on the top contains reports for relevant standards: **US Government**, **ISO**, **PCI**, **SOC**, **Industry & Regional**, and **HITRUST**. You can search for a report and use filters to narrow down the list of reports. Identify the report you want to do view and click on the ↓**Download** link next to a report title.

10. Read the **Privacy Notice** pop-up message and click **Download** to save the report.

11. Close the **Audit reports** blade to return to the **Regulatory compliance** blade.

12. On the top menu, click **Compliance over time workbook**.

13. Before displaying the workbook details, you must configure continuous export to a Log Analytics Workspace on a subscription. To configure Defender for Cloud continuous export to a Log Analytics workspace; refer to the recipes in *Chapter 3, Workflow Automation and Continuous Export*.

14. From the menus, select a **Workspace** option, a **Subscription** option, and a regulatory standard option from the **Standard name** dropdown. The **Regulatory compliance overview** and **Regulatory compliance passed controls over time (weekly)** dashboards display the following:

Figure 6.19 – The Compliance Over Time workbook dashboard

15. On **Regulatory compliance overview**, click on a standard to display the changes for the selected standard and its categorized controls – the passed control percentages, and the 7- and 30-day changes:

Figure 6.20 – The changes over time dashboard

16. Click on a **Main control** category to display its sub-control details: **Current failed assessments, Failed assessments last week,** and **Failed assessments last month.**

17. Furthermore, you can select a sub-control's compliance control and see its assessments over time.

'PA.1' assessments over time

Recommendation name	Current failed resources	Failed resources last week	Failed resources last month
There should be more than one owner assigned to subscriptions	1	1	N/A
External accounts with owner permissions should be removed from your su...	0	0	N/A
Deprecated accounts with owner permissions should be removed from your...	0	0	N/A
A maximum of 3 owners should be designated for subscriptions	0	0	N/A

Figure 6.21 – The assessments over time dashboard

18. Click on a recommendation name to see the current failing resources and the failed resources over time for each subscription.

How it works...

When you configure a continued export for a subscription, regulatory compliance data is stored in the Log Analytics workspace. The **Compliance Over Time** dashboard uses saved data to graphically present and display saved data in more human-friendly and easily understandable color-coded lists and charts.

7

Microsoft Defender for Cloud Workload Protection

Microsoft Defender for Cloud is enabled on all Azure subscriptions by default, while enhanced security features are only available as part of Microsoft Defender for Cloud plans. In this chapter, you will learn how to manage enhanced security and cloud workload protection features available in Defender for Cloud plans, such as vulnerability assessments, **Just-in-Time** (**JIT**) access, adaptive application control, file integrity monitoring, and adaptive network hardening. These workload protection features further extend the protection of Microsoft Defender for Cloud to include important infrastructure and workload security standards, significantly raising the security posture not only of monitored resources but also of the entire cloud environment.

We will cover the following recipes in this chapter:

- Enabling a vulnerability assessment solution
- Enabling and configuring JIT access on a virtual machine
- Requesting access to a JIT-enabled virtual machine
- Configuring the adaptive application control group

- Managing adaptive network hardening

- Remediating vulnerabilities in Azure Container Registry images

- Managing a SQL vulnerability assessment

- Managing file integrity monitoring

Technical requirements

To successfully complete the recipes in this chapter, the following are required:

- An Azure subscription.

- A web browser, preferably Microsoft Edge.

- Microsoft Defender for Cloud plans.

- Resources in an Azure subscription, such as virtual machines, storage, SQL Server, and Logic Apps. Microsoft Defender for Cloud will create resource recommendations based on the resources available.

The code samples can be found at `https://github.com/PacktPublishing/Microsoft-Defender-for-Cloud-Cookbook`.

Enabling a vulnerability assessment solution

Before Microsoft Defender for Cloud can receive vulnerability assessment findings from virtual machines, you must first enable a vulnerability assessment solution on target resources. Once a vulnerability assessment solution has been enabled on a resource, it reports vulnerabilities to Microsoft Defender for Cloud.

In this recipe, you will learn how to automatically enable a vulnerability solution at a subscription level, and how to enable a vulnerability solution on a virtual machine using a recommendation.

Getting ready

Open a web browser and navigate to `https://portal.azure.com`.

How to do it...

First, you will enable automatic deployment of the vulnerability assessment tool and then remediate a recommendation to install a vulnerability assessment solution on a virtual machine. To perform these tasks, complete the following steps:

1. In the Azure portal, open **Microsoft Defender for Cloud**.

2. On the left-hand menu, click **Environment settings**.

3. Click on an arrow next to a management group to display a list of associated subscriptions. Select the desired subscription.

4. From the left-hand menu, click **Auto provisioning**.

5. Set the status of **Vulnerability assessment for machines** to **On**.

6. Click on **Edit configuration** for **Vulnerability assessment for machines**.

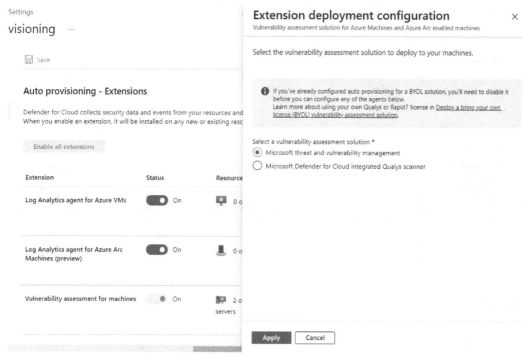

Figure 7.1 – Extension deployment configuration

7. Select the desired vulnerability assessment solution: the **Microsoft threat and vulnerability management** tool or **Microsoft Defender for Cloud integrated Qualys scanner**.

8. Click **Apply** to save the selection.

9. Close the **Auto provisioning - Extensions** page.

    You have enabled the automatic deployment of a vulnerability assessment solution. You will now remediate a recommendation to install a vulnerability assessment solution on a virtual machine or, in other words, manually install a vulnerability tool on a virtual machine.

10. On the menu, select **Recommendations**.

11. On the list of recommendations, expand the **Remediate vulnerabilities** control.

12. Select **Machines should have a vulnerability assessment solution**.

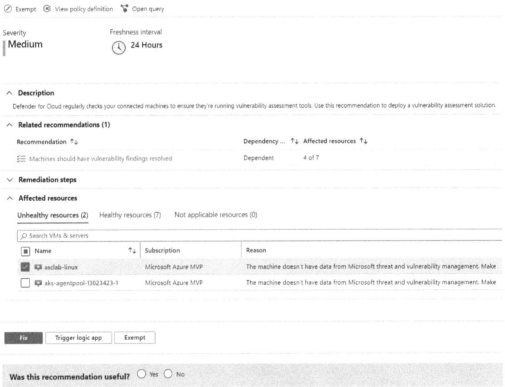

Figure 7.2 – Vulnerability remediation

13. Under the **Affected resources** section, you will see a list of virtual machines that have been identified as **Unhealthy resources**. Click the checkbox next to a virtual machine to install a vulnerability assessment tool.

14. There are three buttons available. Click **Exempt** to exclude the selected virtual machine from future recommendations. Click **Trigger logic app** to run and execute a logic app.

Then, click **Fix**.

> **Tip**
>
> See the *Automating security center recommendations responses* recipe in *Chapter 3, Workflow Automation and Continuous Export*, and the *Creating a recommendation exemption rule* recipe in *Chapter 4, Secure Score and Recommendations*.

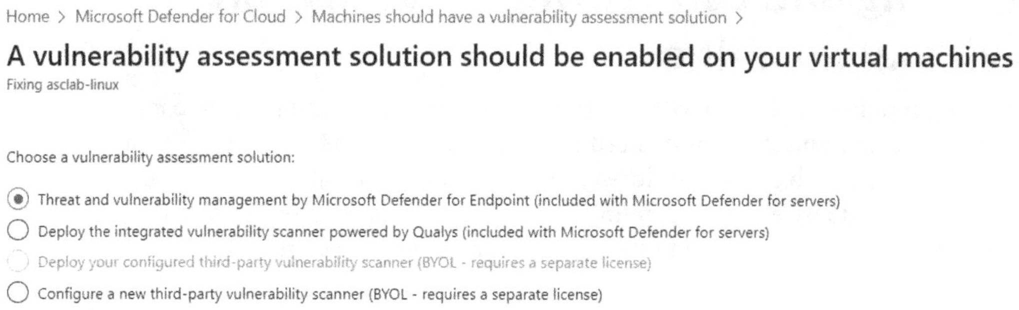

Home > Microsoft Defender for Cloud > Machines should have a vulnerability assessment solution >

A vulnerability assessment solution should be enabled on your virtual machines
Fixing asclab-linux

Choose a vulnerability assessment solution:

◉ Threat and vulnerability management by Microsoft Defender for Endpoint (included with Microsoft Defender for servers)

◯ Deploy the integrated vulnerability scanner powered by Qualys (included with Microsoft Defender for servers)

◯ Deploy your configured third-party vulnerability scanner (BYOL - requires a separate license)

◯ Configure a new third-party vulnerability scanner (BYOL - requires a separate license)

Figure 7.3 – Choosing a vulnerability assessment solution

15. Select **Threat and vulnerability management by Microsoft Defender for Endpoint (included with Microsoft Defender for servers)**.

Optionally, you can select any of the three other available options to install a third-party, or non-Microsoft, vulnerability assessment solution.

16. Click **Proceed**.

17. On the **Fixing resources** blade, review the list of resources that will be onboarded to Microsoft threat and vulnerability management. Click **Fix 1 resource**.

If you selected more than one resource to fix, the button text will change accordingly.

How it works...

Managing endpoint security in a large and dynamic environment can be a difficult task – not all endpoints, that is, physical or virtual machines, might have an endpoint security solution installed or enabled by default. Microsoft Defender for Cloud plans include Microsoft Defender for Endpoint licenses, enabling native protection of Windows and Linux machines hosted in Azure, AWS, and on-premises locations. Microsoft Defender for Cloud endpoint protection includes vulnerability assessment, post-breach detection, threat intelligence, and, what is most important, automatic Defender for Endpoint onboarding on all protected and supported workloads.

Enabling and configuring JIT access on a virtual machine

Operating systems, in our case virtual machines, use networks and ports to communicate. To establish communication, ports need to be accessible. Open, accessible ports are a serious security problem, while closed ports make communication impossible. The problem is, how do we enable communication while staying secure? To solve the problem and to have the best of both solutions, JIT virtual machine access makes ports accessible per request, for a predefined time range.

In this recipe, you will learn how to enable and configure JIT virtual machine access.

Getting ready

Open a web browser and navigate to `https://portal.azure.com`.

How to do it...

To enable and configure JIT virtual machine access, complete the following steps:

1. In the Azure portal, open **Microsoft Defender for Cloud**.
2. On the left-hand menu, click **Workload protections**.

3.   Under the **Advanced protection** section, click on the **Just-in-time VM access** box:

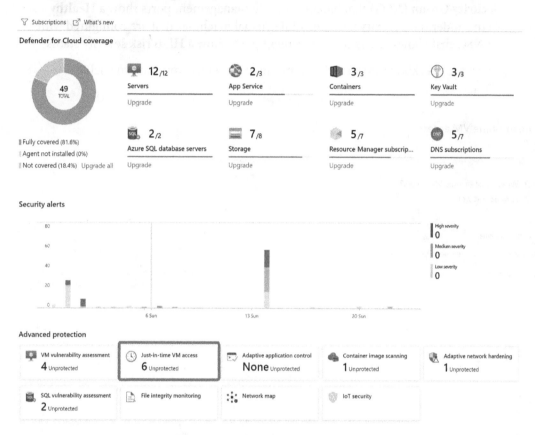

Figure 7.4 – Defender for Cloud coverage

4.   The **Just-in-time VM access** blade contains three tabs, displaying a list of virtual machines for which JIT is already **configured**, a list of virtual machines **not configured** with JIT access, and a list of virtual machines that do not support JIT (**unsupported**).

Click on the **Not Configured** tab.

5.   Virtual machines that have no public IP address or are protected by a **Network Security Group** (**NSG**) that blocks access to management ports show a **Healthy** status under the **Severity** column, while virtual machines that are configured with an NSG that allows access to management ports show a **High** risk severity status.

Select the checkbox next to a virtual machine for which you want to enable JIT access.

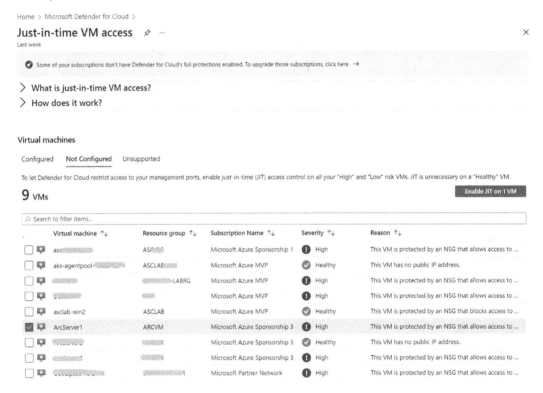

Figure 7.5 – Just-in-time VM access

Note that you can select multiple virtual machines and configure JIT on multiple virtual machines simultaneously. Click the **Enable JIT on 1 VM** button.

Home > Microsoft Defender for Cloud > Just-in-time VM access >

JIT VM access configuration   ⋯   ✕

ArcServer1

+ Add   💾 Save   ✕ Discard

Configure the ports for which the just-in-time VM access will be applicable

Port	Protocol	Allowed source IPs	IP range	Time range (hours)	
22 *(Recommended)*	Any	Per request	N/A	3 hours	•••
3389 *(Recommended)*	Any	Per request	N/A	3 hours	•••
5985 *(Recommended)*	Any	Per request	N/A	3 hours	•••
5986 *(Recommended)*	Any	Per request	N/A	3 hours	•••

Figure 7.6 – JIT VM access configuration

6. The **JIT VM access configuration** blade contains a list of predefined, recommended ports to which JIT VM access will apply: 22, 3389, 5985, and 5986.

> **Note**
>
> By default, JIT VM access configuration includes port 22 (the Secure Shell protocol, or SSH), port 3389 (ms-wbt-server or **Remote Desktop Protocol (RDP)**), and ports 5585 and 5586 (Windows Remote Management service WinRM-HTTP and WinRM-HTTPS or Windows PowerShell default ports).
>
> To change the configured ports' settings, click on a port entry, make the desired changes, and then click the **OK** button.

7. To add a port configuration, click **+ Add**.

8. In the **Port** field, enter the port number.

9. Port configuration can apply to TCP, UDP, or both protocols. Leave the default setting – **Any**.

10. **Allowed source IPs** allows you to configure which source IP addresses or address ranges the JIT settings will apply to. You can choose to allow an IP address allocated at the time of the request, or you can allow multiple IP addresses and IP address blocks, written in a CIDR notation and separated by commas.

Figure 7.7 – Add port configuration

11. Move the slider or enter a number to allow access at the requested time, between 1 and 24 hours.

12. Click **OK**. The port you configured for JIT VM access is added to the list of configured ports, and the virtual machine you configured for JIT access is listed on the **Configured** tab.

How it works...

Obviously, to be able to access virtual machines hosted in Azure and AWS, virtual machines' NSGs must allow traffic to these resources. Without a JIT access feature, administrators would have to spend a tremendous amount of time accommodating access requests – allowing access first, and then denying access once access is no longer needed. The JIT access feature ensures that all these actions are done automatically, in a timely manner, and ensures that no ports are left open, thereby increasing the security of protected workloads.

Requesting access to a JIT-enabled virtual machine

After you have enabled and configured JIT virtual machine access, you can request access to a JIT-enabled virtual machine in multiple ways.

In this recipe, you will learn how to request access to a JIT-enabled virtual machine from Microsoft Defender for Cloud and Azure Virtual Machines.

Getting ready

Open a web browser and navigate to `https://portal.azure.com`.

How to do it...

To request access to a JIT-enabled virtual machine, complete the following steps:

1. In the Azure portal, open **Microsoft Defender for Cloud**.
2. On the left-hand menu, click **Workload protections**.
3. Under the **Advanced protection** section, click on the **Just-in-time VM access** box.
4. Click the **Configured** tab to display a list of virtual machines for which JIT access is configured.
5. Select the checkboxes for the virtual machines for which you want to request JIT access.

Virtual machines

Configured   Not Configured   Unsupported

VMs for which the just-in-time VM access control is already in place. Presented data is for the last week.

3 VMs   Request access

	Virtual machine ↑↓	Approved ↑↓	Last access ↑↓	Connection details	Last user ↑↓
☑ 🖥	asclab-win	0 Requests	N/A	🛡 -	N/A
☑ 🖥	asclab-linux	0 Requests	N/A	🛡 -	N/A
☑ 🖥	ArcServer1	0 Requests	N/A	🛡 -	N/A

Figure 7.8 – JIT on virtual machines

6. Click **Request access**.

7. The **Request access** blade opens, containing a list of preconfigured JIT settings for each virtual machine you selected.

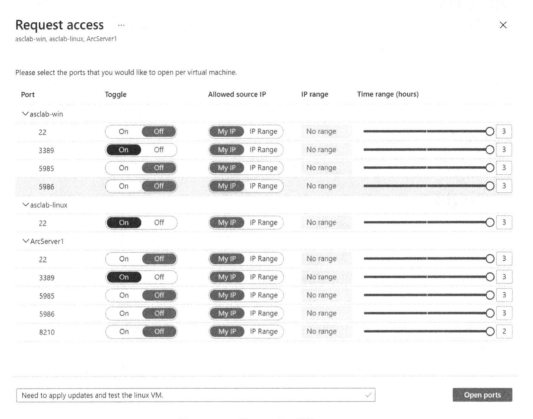

Figure 7.9 – Requesting JIT access

8. Toggle the switches for the ports you want to open, adjusting the time range and allowed source IP addresses.

9. Optionally, type `request justification` and click **Open ports**.

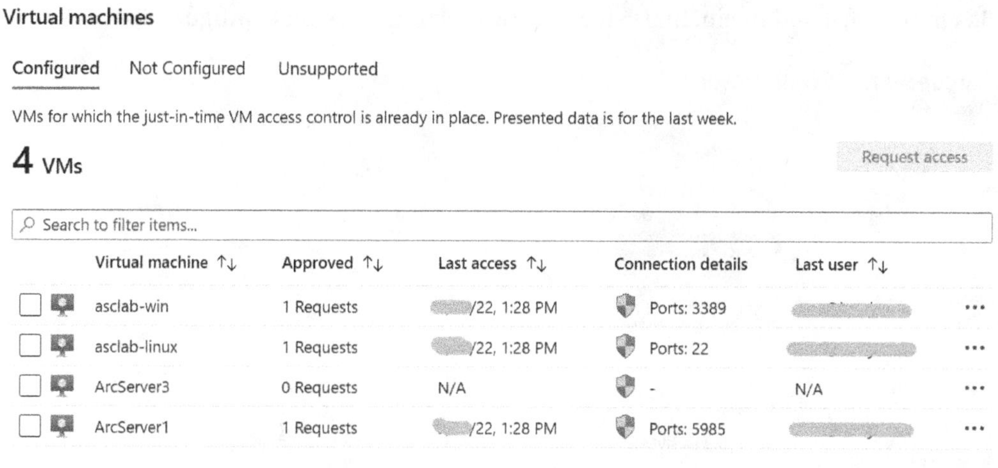

Virtual machines

Configured   Not Configured   Unsupported

VMs for which the just-in-time VM access control is already in place. Presented data is for the last week.

4 VMs

Request access

🔍 Search to filter items...

	Virtual machine ↑↓	Approved ↑↓	Last access ↑↓	Connection details	Last user ↑↓	
☐ 🖥	asclab-win	1 Requests	/22, 1:28 PM	🛡 Ports: 3389		•••
☐ 🖥	asclab-linux	1 Requests	/22, 1:28 PM	🛡 Ports: 22		•••
☐ 🖥	ArcServer3	0 Requests	N/A	🛡 -	N/A	•••
☐ 🖥	ArcServer1	1 Requests	/22, 1:28 PM	🛡 Ports: 5985		•••

Figure 7.10 – JIT-configured virtual machines

10. The **Configured** tab shows virtual machines configured for JIT access and detailed information: the number of approved requests, the most recent access time, and the user, as well as the currently opened ports. Select the **Not Configured** tab to display a list of virtual machines that are not configured with JIT access.

    Alternatively, you can request access to a JIT-enabled virtual machine from the **Azure virtual machine** blade. From the **Not Configured** tab, take note of the name of a virtual machine that is not configured with JIT access. In the top-right corner, click **Home**.

11. Open the **Virtual machines** blade.

12. Select the virtual machine you took note of in *step 10*, the one that is not configured with JIT access.

13. On the left-hand menu, under the **Settings** section, click on **Configuration**:

Figure 7.11 – Machine configuration

14. Click on the **Enable just-in-time** button. You have enabled JIT access on a virtual machine using a virtual machine portal.

How it works...

To further increase the security of protected virtual machines, when requesting JIT access, security administrators can specify which ports are going to be opened, and for how long, specifying ports, allowed source IP addresses and IP ranges, and time ranges more granularly, on each virtual machine individually. Having these ad hoc adjustment capabilities, the security of protected workloads is greatly increased.

Configuring the adaptive application control group

Virtual machines characteristically run similar processes and, after the workloads on virtual machines have been set up, these processes do not change or, at least, no processes other than the intended ones are usually active.

Adaptive application control intelligently tracks and analyzes active and running processes on virtual machines, whitelists these applications, and generates alerts if any other unapproved application is executed.

In this recipe, in the first example, you will configure a recommended adaptive application control group. The second example task will describe how to create a custom adaptive application controls group.

Getting ready

Open a web browser and navigate to `https://portal.azure.com`.

How to do it...

To configure a Microsoft Defender for Cloud adaptive application control group, complete the following steps:

1. In the Azure portal, open **Microsoft Defender for Cloud**.

2. On the left-hand menu, click **Workload protections**.

3. Under the **Advanced protection** section, click on the **Adaptive application control** box.

4. The **Adaptive application controls** blade contains three tabs, each grouping virtual machines according to their status: **Configured**, **Recommended**, and **No recommendation**. The **Configured** tab contains a list of virtual machines that have a whitelist already applied. The **Recommended** tab contains a list of virtual machines for which Microsoft Defender for Cloud has a whitelist recommendation, while the **Not recommended** tab has a list of virtual machines that do not have a whitelist applied or do not support whitelisting and placement under the adaptive application control group.

    There are two ways in which you can create an adaptive application control group: either by clicking on the + **Add custom group** menu button or by configuring a recommended adaptive application control group.

In this example, you will configure a recommended adaptive application control group. Click on the **Recommended** tab.

Configured   **Recommended**   No recommendation

Groups of machines for which we recommend applying application controls to define a list of known-safe applications

Group Name	Machines	State	Severity
∨ 🔑 Microsoft Azure MVP	5		
🖳 GROUP1-EU	1	Open - New	High
🖳 REVIEWGROUP3-EU	2	Open - New	High
🖳 REVIEWGROUP5-EU	1	Open - New	High
🖳 REVIEWGROUP8-EU	1	Open - New	High
∨ 🔑 Microsoft Azure Sponsorship 1	1		
🖳 REVIEWGROUP2-EU	1	Open - New	High
∨ 🔑 Microsoft Azure Sponsorship 3	3		
🖳 GROUP2-EU	2	Open - New	High
🖳 GROUP3-EU	1	Open - New	High
∨ 🔑 Microsoft Partner Network	1		
🖳 REVIEWGROUP2-EU	1	Open - New	High

Figure 7.12 – Adaptive application controls

5. Virtual machines for which recommendations are created are grouped by subscriptions. Click on a group name to configure application control rules. The **Configure application control rules** blade opens.

6. You can review and change the selection of virtual machines in the configuration: applications that are frequent on the selected virtual machines and are highly recommended for defining allowlist rules, and applications that are present on the selected virtual machines but are not critical for including them in an allowlist, but require your attention. Make the necessary adjustments by unchecking unwanted or unnecessary virtual machines and applications and then clicking **Audit**:

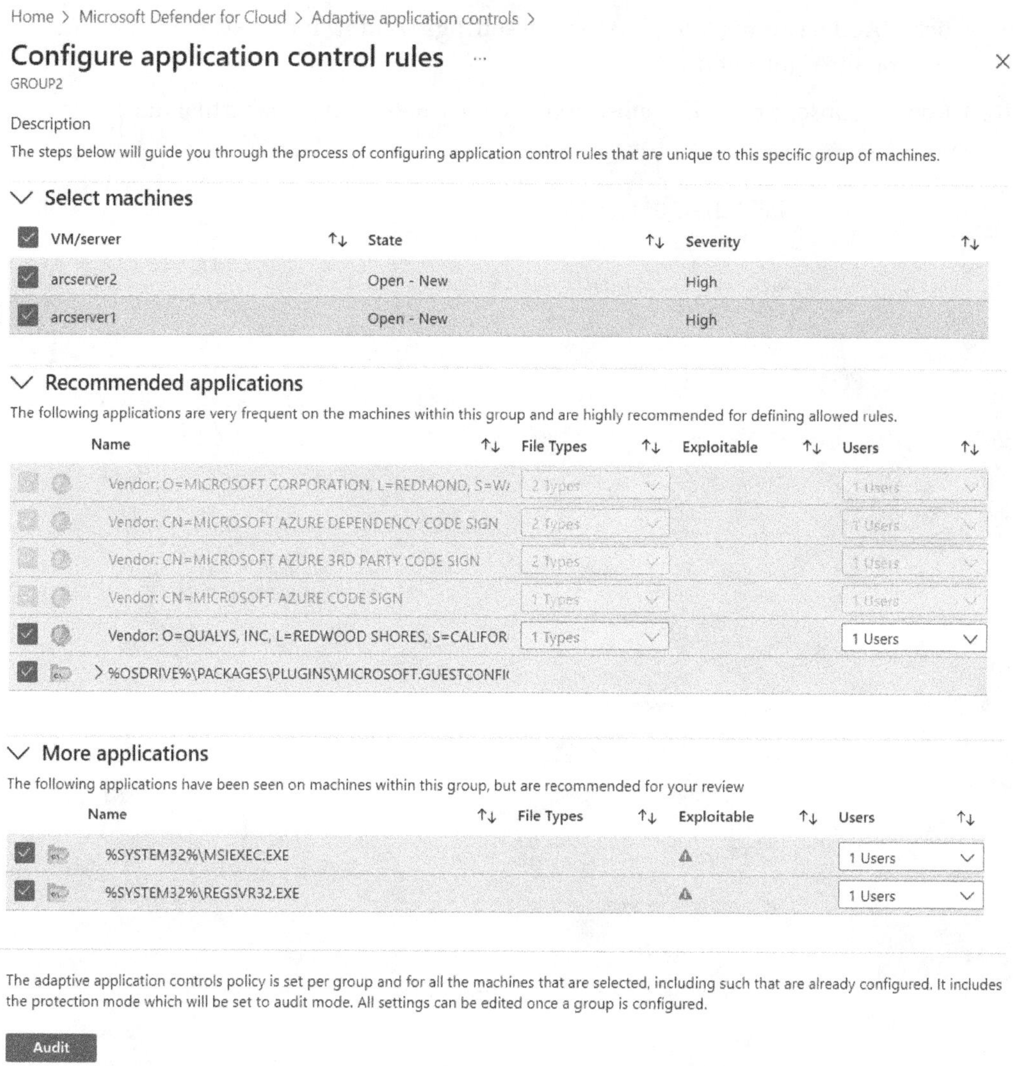

Home > Microsoft Defender for Cloud > Adaptive application controls >

Configure application control rules …   ✕
GROUP2

Description
The steps below will guide you through the process of configuring application control rules that are unique to this specific group of machines.

∨ Select machines

	VM/server	↑↓	State	↑↓	Severity	↑↓
☑	arcserver2		Open - New		High	
☑	arcserver1		Open - New		High	

∨ Recommended applications
The following applications are very frequent on the machines within this group and are highly recommended for defining allowed rules.

	Name	↑↓	File Types	↑↓	Exploitable	↑↓	Users	↑↓
☐ ◎	Vendor: O=MICROSOFT CORPORATION, L=REDMOND, S=W/		2 Types ∨				1 Users ∨	
☐ ◎	Vendor: CN=MICROSOFT AZURE DEPENDENCY CODE SIGN		2 Types ∨				1 Users ∨	
☐ ◎	Vendor: CN=MICROSOFT AZURE 3RD PARTY CODE SIGN		2 Types ∨				1 Users ∨	
☐ ◎	Vendor: CN=MICROSOFT AZURE CODE SIGN		1 Types ∨				1 Users ∨	
☑ ◎	Vendor: O=QUALYS, INC, L=REDWOOD SHORES, S=CALIFOR		1 Types ∨				1 Users ∨	
☑ ▣	> %OSDRIVE%\PACKAGES\PLUGINS\MICROSOFT.GUESTCONFI(							

∨ More applications
The following applications have been seen on machines within this group, but are recommended for your review

	Name	↑↓	File Types	↑↓	Exploitable	↑↓	Users	↑↓
☑ ▣	%SYSTEM32%\MSIEXEC.EXE				⚠		1 Users ∨	
☑ ▣	%SYSTEM32%\REGSVR32.EXE				⚠		1 Users ∨	

The adaptive application controls policy is set per group and for all the machines that are selected, including such that are already configured. It includes the protection mode which will be set to audit mode. All settings can be edited once a group is configured.

[Audit]

Figure 7.13 – Configure application control rules

7. A message, **AAC group configuration request succeeded**, will confirm the configuration. The configured group name will be displayed on the **Adaptive application controls** blade, under the **Configured** tab. If not, refresh the page.

8. The second example task will describe how to create a custom adaptive application controls group. If not already on the **Adaptive application controls** blade, complete *steps 1-3* in this recipe to open it.

9. Click **+ Add custom group**. The **Group settings** window for a new machine group opens on the right-hand side.

10. Choose a subscription. The subscription choice determines the virtual machines you will be able to choose from.

Figure 7.14 – Adaptive application control group settings

11. Select the location, **Global** or **Europe**, or any other location that is relevant and available for your subscription.

12. Type a name for a group (uppercase letters and numbers are allowed), and then choose the operating system type and the workload environment type.

13. Choose the file type protection mode: **EXE** (executable or application), **MSI** (Microsoft Installer), or **SCRIPT**.

14. You can check configured and unconfigured virtual machines to add them to a group.

15. Click **Audit** to finish creating the group.

How it works...

Adaptive application controls are used to define which applications are trusted and safe to run – you use them to make allowlists of applications that are authorized to run on specific workloads. If an application is executed but is not allowed to run, adaptive application control will generate an alert, thereby notifying you of suspicious behavior, and potential malware presence.

Additionally, adaptive application controls help you stay compliant, ensuring only approved, licensed, and current software is used

If you added previously configured virtual machines to the new group, machines will be reconfigured with the rules that are inherited from the new group.

At this time, adaptive application controls provide alerts only, while the option to enforce rules is not available.

Managing adaptive network hardening

Adaptive network hardening does not need any configuration to work – it is an agentless protection feature. Assisted by machine learning algorithms, it monitors virtual network traffic flow to and from resources, considers NSG inbound and outbound rules, analyzes, and makes recommendations to further strengthen NSG rules.

In this recipe, you will learn how to manage adaptive network hardening recommendations in Microsoft Defender for Cloud.

Getting ready

Open a web browser and navigate to `https://portal.azure.com`.

How to do it...

To manage a Microsoft Defender for Cloud default security policy, complete the following steps:

1. In the Azure portal, open **Microsoft Defender for Cloud**.

2. On the left-hand menu, click **Workload protections**.

3. Under the **Advanced protection** section, click on the **Adaptive network hardening** box.

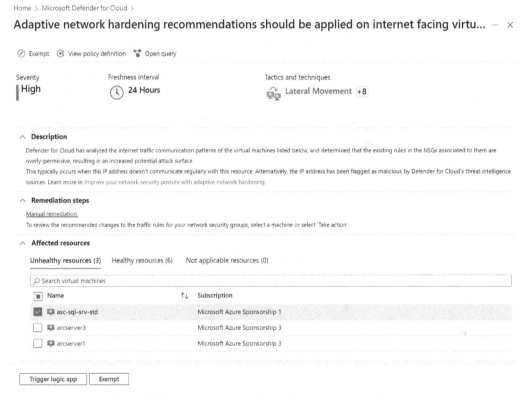

Figure 7.15 – Adaptive network hardening recommendations

4. The **Affected resources** group displays three groups of resources:

    I.   **Unhealthy resources**: Virtual machines that have alerts and recommendations to restrict network traffic

    II. **Healthy resources**: Virtual machines that have no alerts and network restriction recommendations

III. **Not applicable resources**: Virtual machines for which adaptive network hardening algorithms could not be applied because virtual machines are not protected by Microsoft Defender for Servers. There is not enough data (you will need at least 30 days of traffic data) or virtual machines are classic virtual machines (not supported).

Select the **Unhealthy** resources tab. From the list of unhealthy virtual machines, click on a virtual machine name.

5. The **Manage adaptive network hardening recommendations** blade opens.

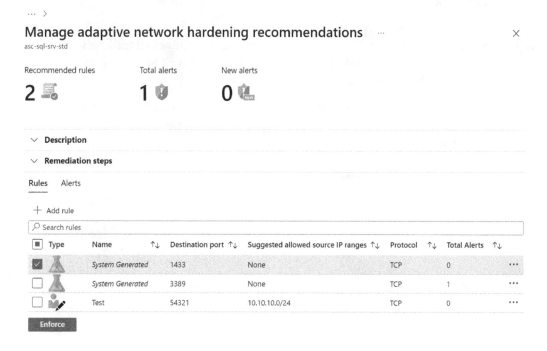

Figure 7.16 – Manage adaptive network hardening recommendations

The **Rules** tab displays the rules that are recommended by adaptive network hardening algorithms, while the **Alerts** tab lists alerts generated on the observed resources, that is, virtual machines, such as **Traffic detected from IP addresses recommended for blocking** alerts. Optionally, to add a new rule, click the **+ Add rule** button, where you need to specify a rule name, destination port, allowed source IP ranges or individual IP addresses, and a protocol.

To proceed, select all the rules and then click **Enforce**.

6.   Adaptive network hardening modifies the NSG (or groups) by adding the rules defined and selected in *step 5*, thus strengthening virtual machine security by restricting network traffic.

How it works...

Based on several aspects that are considered, such as various indicators of compromise, NSG rules, threat intelligence information, virtual network traffic characteristics, and others, adaptive network hardening offers recommendations to further reinforce virtual network traffic rules.

For example, if NSG rules are configured to allow inbound traffic on a specific IP address range – for example, /24 subnet, and on 10 different ports, while actual traffic always uses only a portion of a subnet and only 4 ports – adaptive network hardening suggests restricting traffic to IP addresses and ports that are actually used.

The inbound port rules list before enforcing adaptive network hardening changes shows that RDP and SQL ports (3389 and 1433) are allowed from any source to any destination, as shown in the following screenshot:

Inbound port rules	Outbound port rules	Application security groups		Load balancing			

Network security group asc-sql-srv-std-nsg (attached to network interface: asc-sql-srv-std988)
Impacts 0 subnets, 1 network interfaces

Add inbound port rule

Priority	Name	Port	Protocol	Source	Destination	Action	
300	⚠ RDP	3389	TCP	Any	Any	✓ Allow	...
1500	⚠ default-allow-sql	1433	TCP	Any	Any	✓ Allow	...
2000	Port_49787	49787	Any	Any	Any	✓ Allow	...
2100	Port_49788	49788	Any	Any	Any	✓ Allow	...
2210	Port_49789	49789	Any	Any	Any	✓ Allow	...
65000	AllowVnetInBound	Any	Any	VirtualNetwork	VirtualNetwork	✓ Allow	...
65001	AllowAzureLoadBalancerInBound	Any	Any	AzureLoadBalancer	Any	✓ Allow	...
65500	DenyAllInBound	Any	Any	Any	Any	⊗ Deny	...

Figure 7.17 – Inbound port rules before enforcing adaptive network hardening rules

After enforcing adaptive network hardening rules on the NSG attached to the virtual machine network interface card affected, adaptive network hardening modifies NSG rules and restricts network traffic to include only the allowed ports, protocols, sources, and destinations.

Inbound port rules	Outbound port rules	Application security groups	Load balancing

Network security group asc-sql-srv-std-nsg (attached to network interface: asc-sql-srv-std988)
Impacts 0 subnets, 1 network interfaces

Add inbound port rule

Priority	Name	Port	Protocol	Source	Destination	Action	
298	Test	54321	TCP	10.10.10.0/24	172.16.0.4	Allow	•••
299	SecurityCenter-ANHRule_TCP_Inbound_DENY...	54321,3389,1433	TCP	Internet	172.16.0.4	Deny	•••
300	⚠ RDP	3389	TCP	Any	Any	Allow	•••
1500	⚠ default-allow-sql	1433	TCP	Any	Any	Allow	•••
2000	Port_49787	49787	Any	Any	Any	Allow	•••
2100	Port_49788	49788	Any	Any	Any	Allow	•••
2210	Port_49789	49789	Any	Any	Any	Allow	•••
65000	AllowVnetInBound	Any	Any	VirtualNetwork	VirtualNetwork	Allow	•••
65001	AllowAzureLoadBalancerInBound	Any	Any	AzureLoadBalancer	Any	Allow	•••
65500	DenyAllInBound	Any	Any	Any	Any	Deny	•••

Figure 7.18 – Inbound port rules after enforcing adaptive network hardening rules

There's more...

On the **Adaptive network hardening recommendations** blade, if you select a checkbox next to the name of the virtual machine, you can **Trigger logic app** or **Exempt** a virtual machine or machines from being recommended to restrict network traffic and appearing under the **Unhealthy resources** tab.

Remediating vulnerabilities in Azure Container Registry images

The Microsoft Defender for Containers plan is a part of Microsoft Defender for Cloud responsible for protecting containers. Its container protection includes Kubernetes services in Azure (Azure Kubernetes Service, or AKS), Amazon Kubernetes services, and Azure Arc-enabled Kubernetes clusters.

In this recipe, you will learn how to remediate vulnerabilities in Azure Container Registry images, using Microsoft Defender for Containers vulnerability assessments.

Getting ready

Open a web browser and navigate to `https://portal.azure.com`.

How to do it...

To remediate vulnerabilities in Azure Container Registry images, using Microsoft Defender for Containers vulnerability assessments, complete the following steps:

1. In the Azure portal, open **Microsoft Defender for Cloud**.

2. On the left-hand menu, click **Workload protections**.

3. Under the **Advanced protection** section, click on the **Container image scanning** box.

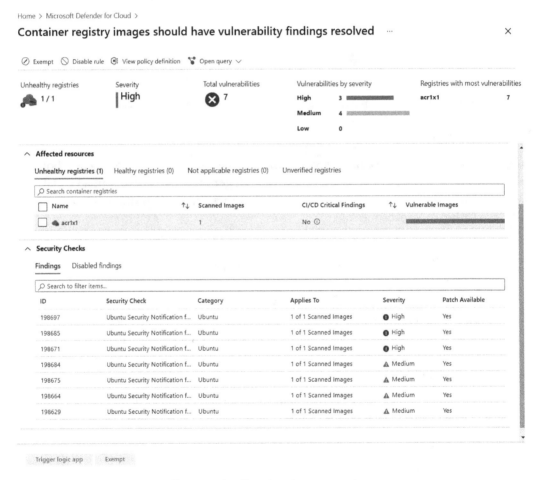

Figure 7.19 – Container image scanning

4. The **Affected resources** section contains the **Unhealthy registries** tab and a list of container registries containing vulnerable images. Other tabs include **Healthy registries**, **Unverified registries**, and **Not applicable registries**. The **Security Checks** section contains a list of all security vulnerability findings.

    To open a container registry health page, under the **Affected resources** section, click on a registry name.

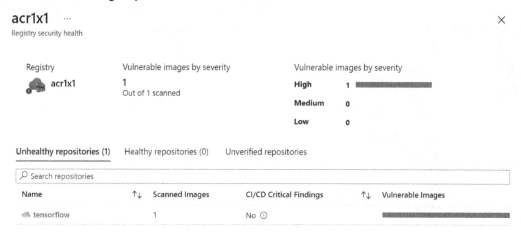

Figure 7.20 – Registry security health

5. The **Registry security health** page covers the health of registry repositories: vulnerable images by severity, unhealthy repositories, healthy repositories, and unverified repositories. Select the **Unhealthy repositories** tab and click on a repository name.

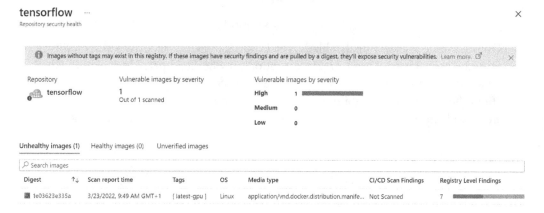

Figure 7.21 – Repository security health

6. Similarly, the **Repository security health** blade gathers images' health information, such as the last scan time, operating system type, and registry-level vulnerability findings. Select the **Unhealthy images** tab and click on an image name.

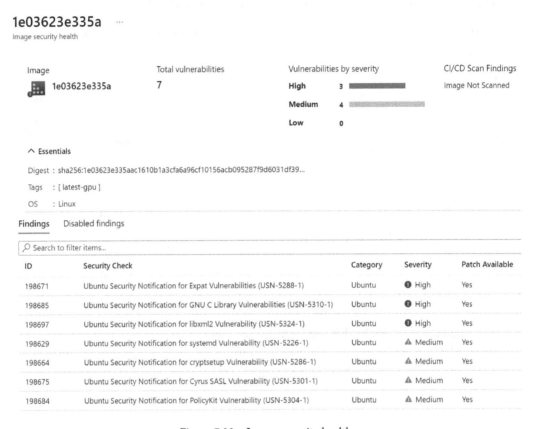

Figure 7.22 – Image security health

7. On the **Image security health** blade, the **Findings** tab lists security vulnerability findings, their severity level, and information on whether a patch is available. Click on a security vulnerability check to open a blade containing detailed information about the security vulnerability: description, general information, remediation steps, and any additional information, as well as any other affected resources.

Figure 7.23 – Image security finding

8. Under the **Remediation** or **Additional information** sections, click a link. A new
    page opens in a web browser containing information about a vulnerability and
    instructions on how to download an update package and update instructions to
    fix a vulnerability.

How it works...

The Microsoft Defender for Containers plan comprises a vulnerability assessment of images stored in Azure Container Registry and running in AKS, threat protection for nodes and clusters, and environment hardening, and provides misconfiguration information and threat mitigation guidelines.

Managing a SQL vulnerability assessment

Microsoft Defender plans include two Microsoft Defender for SQL protection options: **Azure SQL Databases**, which includes single databases and elastic pools, Azure SQL Managed Instance, and Azure Synapse Analytics, and **Azure SQL server on machines**, which applies to SQL on Azure Virtual Machines, SQL servers on-premises, and Azure Arc-enabled SQL servers.

In this recipe, you will learn how to manage SQL vulnerability assessments in Microsoft Defender for Cloud.

Getting ready

Open a web browser and navigate to `https://portal.azure.com`.

How to do it...

To manage SQL vulnerability assessments, complete the following steps:

1. In the Azure portal, open **Microsoft Defender for Cloud**.
2. On the left-hand menu, click **Workload protections**.
3. Under the **Advanced protection** section, click on the **SQL vulnerability assessment** box.
4. A **SQL databases should have vulnerability findings resolved** blade opens. The top menu contains standard buttons that allow you to exempt a recommendation from any scope so that it doesn't affect your secure score, disable a rule to disable one or more findings for this recommendation, view a policy definition, or open a query that will return affected resources or security findings.

Home > Microsoft Defender for Cloud >

SQL databases should have vulnerability findings resolved   ...

⊘ Exempt   ⊘ Disable rule   ⊛ View policy definition   ⚑ Open query ⌄

ⓘ SQL Vulnerability Assessment rules have been updated. This may impact your scan results. Learn more →

Unhealthy servers	Total findings	Findings by severity			Servers with most findings	
🗄 2 / 2	✖ 8	High	2	▬▬▬▬	asc-db-srv	5
		Medium	4	▬▬▬▬▬▬▬	skkloudatechdb	3
		Low	2	▬▬▬		

∧ **Description**

SQL Vulnerability assessment scans your database for security vulnerabilities, and exposes any deviations from best practices such as misconfigurations, excessive permissions, and unprotected sensitive data. Resolving the vulnerabilities found can greatly improve your database security posture. Learn more

∧ **Affected resources**

Unhealthy resources (2)   Healthy resources (0)   Not applicable resources (0)

🔎 Search databases			
☐ Name		Subscription	Failed Checks
☐ 🗄 asc-db-srv	↑↓	Microsoft Azure Sponsorship 1	5 in 4 unhealthy databases
☐ 🗄 skkloudatechdb		Microsoft Azure MVP	3 in 2 unhealthy databases

∧ **Security Checks**

Findings   Passed   Disabled findings

Benchmarks: All					⌄
🔎 Search to filter items...					
ID	Security check	Category	Applies to	Benchmark	Severity
VA2065	Server-level firewall rules should be tracked and ...	Surface Area Reduction	2 of 2 databases		ⓘ High
VA1143	'dbo' user should not be used for normal service ...	Surface Area Reduction	4 of 4 databases	FedRAMP	⚠ Medium
VA2130	Track all users with access to the database	Authentication And Au...	2 of 6 databases	SOX	ⓘ Low

Trigger logic app	Exempt

Figure 7.24 – SQL vulnerability assessment

5. The **Security Checks** section contains current security findings, as well security findings that have already passed and that are disabled. Click on the **Findings** tab and select a security finding.

6. A security finding information blade opens, containing detailed information about a vulnerability, remediation steps, the security impact of the vulnerability, and affected and dismissed resources. After reviewing the information, close the blade.

7.   The **Affected resources** tab contains a list of unhealthy, healthy, and not applicable resources. Select the **Unhealthy resources** tab and click on a database.

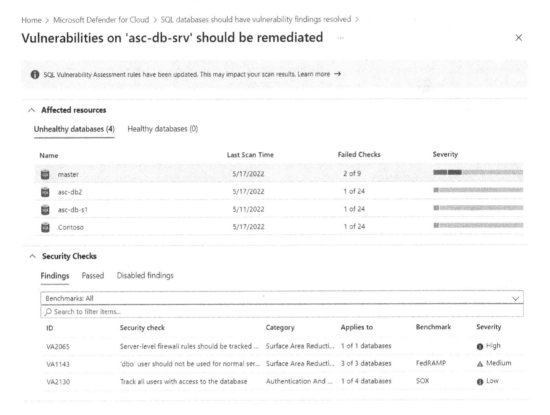

Figure 7.25 – Vulnerabilities that should be remediated

8. The **Vulnerabilities on '<sqlservername>' should be remediated** blade opens. The
    Unhealthy databases tab contains the list of databases that failed security checks,
    and the **Security Checks** tab lists applicable security vulnerability findings. Click on
    a database name to open the database security vulnerability findings blade.

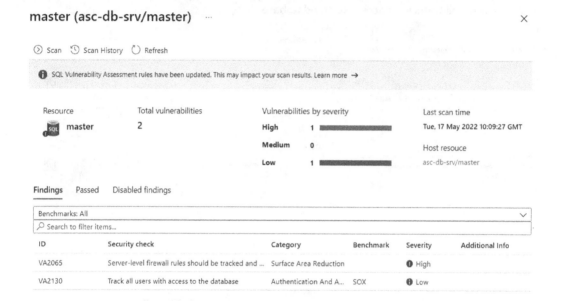

Figure 7.26 – Database security vulnerability findings

9. The database security vulnerability findings blade includes a summary of total
    vulnerabilities, vulnerabilities by severity, and the last scan time, as well as a list of
    security vulnerability findings that includes current and past findings, as well as an
    option to disable findings.

On the top menu, click **Scan** to execute a vulnerability scan, and **Scan history** to display the history of scans executed against the current database. Click on a vulnerability finding to see its details and to remediate it.

Figure 7.27 – SQL vulnerability finding remediation

10. In this example, to remediate the **VA2130 - Track all users with access to the database** vulnerability assessment, you need to revoke unnecessary access to a database and add allowed access to the baseline. To accept the results as a baseline as being acceptable for this database or environment, click **Add all results as baseline**. At the **Set baseline** prompt, click **Yes** to approve the changes to the baseline. As the baseline changes, you need to run a new scan to see the updated results.

11. Close the vulnerability assessment blade to return to the previous blade.

12. On the database blade, on the top menu, click **Scan** to execute a new vulnerability scan.

13. Wait for a scan to finish and click **Scan History**. You should see new, updated results.

14. Close the scan history and click **Refresh** to refresh the database findings and passed results.

15. Click on the **Passed** tab. Identify the **Additional Info** column for the vulnerability you have just remediated and change the baseline. Observe the **Pass Per Baseline** information, confirming the changes to the baseline.

How it works...

Microsoft Defender for Cloud offers a range of SQL protection capabilities, enabling safeguarding databases, detecting anomalous activities, and discovering and mitigating database vulnerabilities, in addition to advanced threat protection and SQL vulnerability assessment capabilities – all through a single *pane of glass*.

To get the advantage of SQL protection capabilities in Microsoft Defender for Cloud, you must enable Microsoft Defender database SQL protection plans.

Managing file integrity monitoring

Microsoft file integrity monitoring, or FIM, is a part of Microsoft Defender for Cloud that enables change monitoring of files, Linux system files, Windows registries, application software, operating system files, and other file-level changes that might signal an attack. In this recipe, you will learn how to manage FIM in Microsoft Defender for Cloud.

Getting ready

Open a web browser and navigate to `https://portal.azure.com`.

How to do it...

To manage FIM in Microsoft Defender for Cloud, complete the following steps:

1. In the Azure portal, open **Microsoft Defender for Cloud**.
2. On the left-hand menu, click **Workload protections**.

3.   Under the **Advanced protection** section, click on the **File Integrity Monitoring** box.

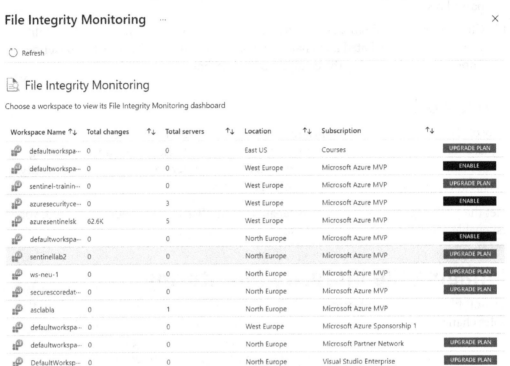

Figure 7.28 – File Integrity Monitoring

4.   FIM requires Microsoft Defender for Servers. Refer to *Chapter 1, Getting Started with Microsoft Defender for Cloud*, for more information on enabling Defender for Cloud plans. If a Log Analytics workspace does not have a Defender for Servers plan enabled, click **Upgrade plan**. Otherwise, click **Enable**. The **Enable File Integrity Monitoring** blade opens.

Home > File Integrity Monitoring >

Enable File Integrity Monitoring   ...   ✕
azuresecuritycentersk

∧   What is File Integrity Monitoring?

File Integrity Monitoring (FIM), also known as change monitoring, validates files and registries integrity of operating system, application software, and others for changes that might indicate an attack. A comparison method is used to determine if the current checksum of the file is different from the latest scan of the file. You can leverage this comparison to determine if valid or suspicious modifications have been made to your files.

ⓘ Enabling file integrity monitoring affects all machines connected to the selected workspace (azuresecuritycentersk)

Windows Servers	Linux Servers	LEARN MORE
3	**0**	Learn more about File Integrity Monitoring ☐

Recommended settings

☑ > Windows Files

☑ > Registry

☑ > Linux Files

ⓘ File Integrity Monitoring (FIM) uploads data to the Log Analytics workspace. Data charges will apply, based on the amount of data you upload. To learn more about Log Analytics pricing click here.

Selected settings from above are applied. You can modify the settings later using 'File Integrity Monitoring' settings

File Integrity Monitoring leverages the Change Tracking solution enabled on your workspace.

Enable File Integrity Monitoring

Figure 7.29 – Enable File Integrity Monitoring

5. Recommended settings for monitoring include Windows files, Windows registry entries, and Linux files. You can make any changes to recommended settings now, or later. To enable FIM on all machines connected to the selected Log Analytics workspace, click **Enable File Integrity Monitoring**.

6.   Once deployment is completed, return to the **File Integrity Monitoring** blade, and select a workspace that has FIM enabled.

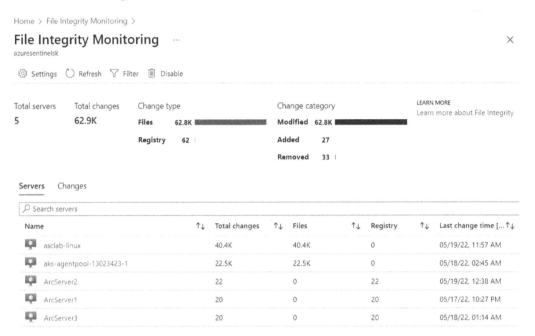

Figure 7.30 – File Integrity Monitoring on a selected Log Analytics workspace

7.   The FIM for a selected Log Analytics workspace displays a monitoring summary, such as total servers and total changes, change types, and category values. The tabs list displays monitored machines and changes that have been made to monitored entities. For additional FIM configuration, on the top menu, click **Settings**.

8.   The **Workspace Configuration** blade allows you to manage change tracking for the Windows registry, Windows files, and Linux files. You can enable or disable change tracking for listed files and registry keys, and specify your own files and registry keys to track changes. Click the **File Content** tab.

Figure 7.31 – Workspace Configuration

9.   Moreover, the **File Content** tab allows you to select and link a storage account that will store file content, to enable the viewing of file content before and after the change. Click the **Link** button.

10. Select a subscription and a storage account. Ensure **Upload file content for all settings** is **On** to enable file content upload for all currently tracked files. Click **Save**.

11. Click the **Windows Services** tab. Move the slider to adjust the collection frequency for Windows Services changes between the allowed values of 10 seconds and 30 minutes.

How it works...

File integrity monitoring uses the Log Analytics agent to upload data to a Log Analytics workspace. FIM compares the state of monitoring entities during a previous scan with the current state and signals whether any changes occurred between the two scans and whether any questionable changes were made.

8
Firewall Manager

In this chapter, you will learn how to secure Azure assets and public endpoints by controlling network traffic to and from Azure.

Microsoft Defender for Cloud is a security monitoring platform that unifies infrastructure monitoring and advanced threat protection, while getting you secured faster, protecting you against threats, and strengthening the security posture of your workloads.

The Microsoft Defender for Cloud Overview dashboard shows six central tiles – **Secure Score**, **Regulatory Compliance**, **Workload protections**, **Inventory**, **Information protection**, and **Firewall Manager**.

The **Firewall Manager** tile shows the status of networks and hubs and, while separate products, Azure Firewall and Firewall Manager are crucial network protection services that safeguard Azure assets.

The prominent Firewall Manager position in Microsoft Defender for Cloud, and its importance in protecting Azure workloads, requires knowing how to operate Firewall Manager and deploy an Azure firewall, Azure firewall policies, secured networks, and hubs.

> **Note**
> At the time of writing, Azure Firewall Premium was still in preview, so changes as regards product functionality and features, as well as Azure portal changes, are possible.

We will cover the following recipes in this chapter:

- Creating an Azure firewall
- Creating an Azure firewall using PowerShell
- Creating an Azure firewall policy
- Creating an Azure firewall policy using PowerShell

Technical requirements

To complete the recipes in this chapter, the following are required:

- An Azure subscription
- Azure PowerShell
- A web browser, preferably Microsoft Edge

The code samples can be found at `https://github.com/PacktPublishing/Microsoft-Defender-for-Cloud-Cookbook`.

Creating an Azure firewall

An Azure firewall is a managed **Platform-as-a-Service (PaaS)** solution that protects resources residing on **Azure Virtual Network**. The Microsoft Defender for Cloud Overview page supports and displays the status of **Firewall Manager** and its supported services, firewalls, and hubs. In this recipe, you will create an **Azure Firewall Standard SKU**.

Getting ready

In these examples, you can choose your own user-defined values instead of the examples provided.

To get ready for an Azure firewall deployment and to complete the preliminary steps, perform the following steps:

1. Open a web browser and navigate to `https://portal.azure.com`.
2. Open **Virtual Networks** and click + **Create**. The **Create Virtual Network** blade opens.
3. In the **Create Virtual Network** blade, under the **Basics** tab, choose **Azure Subscription**, and, under **Resource Group**, click **Create new**, and then type `Firewall` for a resource group name.

4. Under the **Instance details** group, in the **Name** field, type `PacktPublishing` for the **Virtual Network** name.

5. Choose **West Europe** for an Azure Region.

6. Then, click **IP Addresses**.

7. For **IPv4 address space**, delete any existing address space ranges and type `172.16.0.0/16`.

8. Under **Subnet name**, click on **Default**.

9. Change the subnet name to **Production**. For **Subnet address range**, type `172.16.0.0/24`. Click **Save**.

10. Click **+ Add subnet**. In the **Subnet name** field, type `AzureFirewallSubnet`.

> **Note**
>
> This subnet will host an Azure firewall instance and the subnet name *must* be `AzureFirewallSubnet`, which is a reserved subnet name.

11. For **Subnet address range** type `172.16.1.0/26`. Click **Save**.

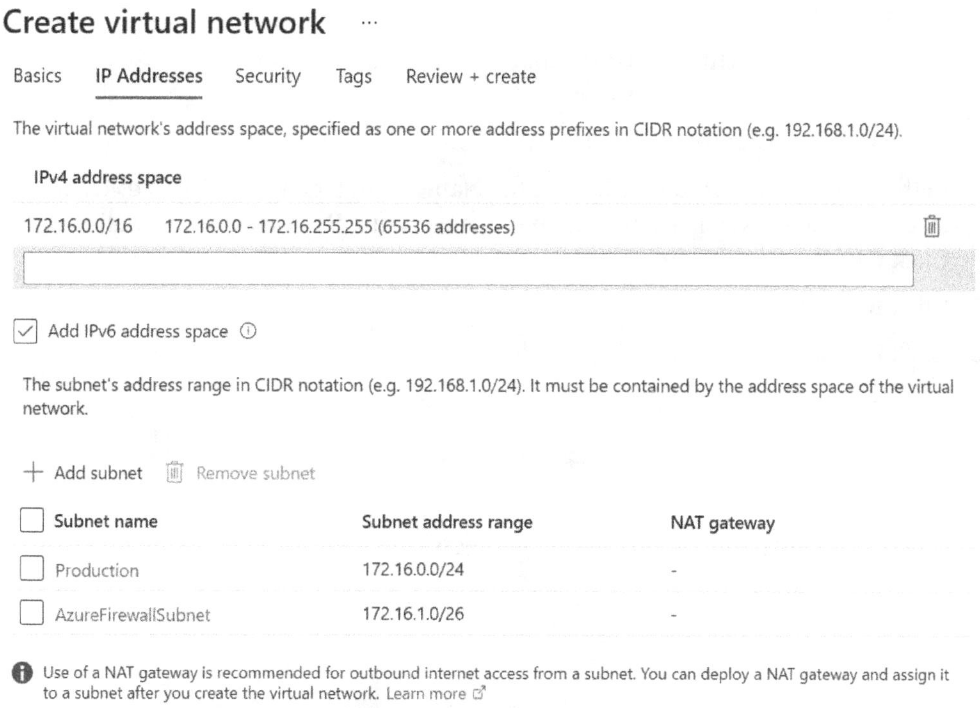

Create virtual network   ⋯

Basics   **IP Addresses**   Security   Tags   Review + create

The virtual network's address space, specified as one or more address prefixes in CIDR notation (e.g. 192.168.1.0/24).

IPv4 address space

172.16.0.0/16	172.16.0.0 - 172.16.255.255 (65536 addresses)	🗑

☑ Add IPv6 address space ⓘ

The subnet's address range in CIDR notation (e.g. 192.168.1.0/24). It must be contained by the address space of the virtual network.

+ Add subnet   🗑 Remove subnet

☐ Subnet name	Subnet address range	NAT gateway
☐ Production	172.16.0.0/24	-
☐ AzureFirewallSubnet	172.16.1.0/26	-

ⓘ Use of a NAT gateway is recommended for outbound internet access from a subnet. You can deploy a NAT gateway and assign it to a subnet after you create the virtual network. Learn more ⟳

Figure 8.1 – Create virtual network

12. Click **Review + create**.

13. Click **Create**.

How to do it...

To create a **Standard tier** Azure firewall, complete the following steps:

1. In the Azure portal, open **Firewalls**.

2. Click **+ Create**.

3. Choose **Azure Subscription** and, in **Resource Group**, choose **Firewall**.

4. Under **Instance details**, in the **Name** field, type `Packt-Firewall`.

5. For the Region, select the same Region as the resource group and the virtual network Region that you used in the *Getting ready* section. In this case, choose `West Europe` for an Azure Region.

6. Optionally, choose none, one, or more Availability Zones.

7. For **Firewall tier**, choose **Standard**.

8. Under **Firewall management**, choose **Use Firewall rules (classic) to manage this firewall**.

9. Under **Choose a virtual network**, choose **Use existing**. Choose the **PacktPublishing** virtual network.

10. In the **Public IP address** field, click **Add new**.

11. In the **Add a public IP** dialog box, in the **Name** field, type `Packt-Firewall-PIP`. An IP address will be created in the same Azure Region as the firewall. Click **OK**.

12. Click **Review + create**.

13. Click **Create**:

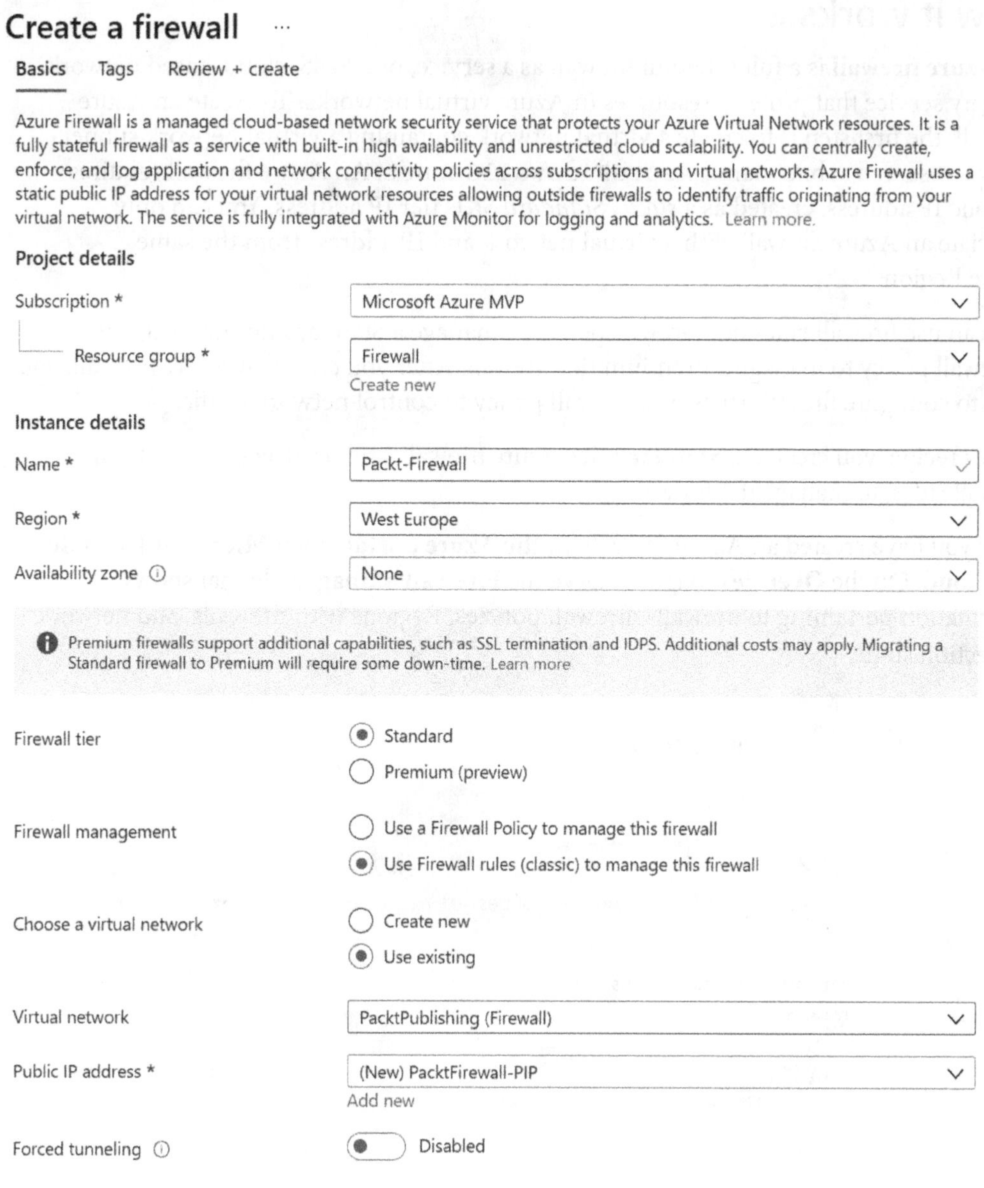

Create a firewall   ...

Basics   Tags   Review + create

Azure Firewall is a managed cloud-based network security service that protects your Azure Virtual Network resources. It is a fully stateful firewall as a service with built-in high availability and unrestricted cloud scalability. You can centrally create, enforce, and log application and network connectivity policies across subscriptions and virtual networks. Azure Firewall uses a static public IP address for your virtual network resources allowing outside firewalls to identify traffic originating from your virtual network. The service is fully integrated with Azure Monitor for logging and analytics.   Learn more.

Project details

Subscription * | Microsoft Azure MVP   ∨ |

    Resource group * | Firewall   ∨ |
                                          Create new

Instance details

Name * | Packt-Firewall   ✓ |

Region * | West Europe   ∨ |

Availability zone ⓘ | None   ∨ |

> ⓘ Premium firewalls support additional capabilities, such as SSL termination and IDPS. Additional costs may apply. Migrating a Standard firewall to Premium will require some down-time. Learn more

Firewall tier   ⦿ Standard
                                         ◯ Premium (preview)

Firewall management   ◯ Use a Firewall Policy to manage this firewall
                                         ⦿ Use Firewall rules (classic) to manage this firewall

Choose a virtual network   ◯ Create new
                                         ⦿ Use existing

Virtual network | PacktPublishing (Firewall)   ∨ |

Public IP address * | (New) PacktFirewall-PIP   ∨ |
                                         Add new

Forced tunneling ⓘ   (⬤   ) Disabled

Figure 8.2 – Create a firewall

How it works...

An **Azure firewall** is a fully stateful firewall as a service, or a PaaS, cloud-based network security service that protects resources in Azure virtual networks. To create an Azure firewall, the first step is to create a virtual network containing a virtual network subnet using a reserved subnet name, `AzureFirewallSubnet`. An Azure firewall requires a public IP address, created as a *static*, *Standard SKU* tier IP address. You can only associate an Azure firewall with a virtual network and IP address from the same Azure Region.

You can use firewall rules or a firewall policy to manage a Standard tier firewall, and a firewall policy to manage a Premium tier firewall. After you create an Azure firewall, you have to configure firewall rules or a firewall policy to control network traffic.

In this recipe, you created a **Standard** tier Azure firewall, and used a classic method, firewall rules, to manage the firewall.

After you have created an Azure firewall, in **the Azure portal**, open **Microsoft Defender for Cloud**. On the **Overview** page, observe the **Firewall Manager** tile that shows information pertaining to firewalls, firewall policies, Regions with firewalls, and network protection status.

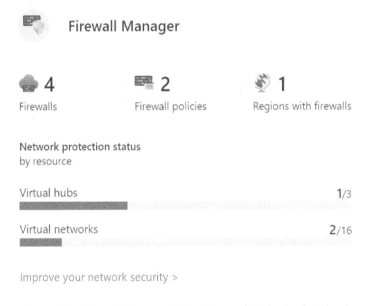

Figure 8.3 – Firewall Manager tile in Microsoft Defender for Cloud

Creating an Azure firewall using PowerShell

In scenarios where interactivity in the Azure portal is minimized and deployments are automated and scripted, creating an Azure firewall using PowerShell is the choice of many Azure administrators. In this recipe, you will create an Azure Firewall Standard SKU using PowerShell.

Getting ready

Open your preferred PowerShell tool – this could be Visual Studio Code, Windows PowerShell ISE, the PowerShell console, or something else.

Sign in to your Azure account: `Connect-AzAccount`.

How to do it...

To create an Azure firewall using PowerShell, complete the following steps:

1. Define resources – resource group, location (Region), virtual network, subnets, IP address, and firewall names:

    ```
    $RGName="Firewall"
    $Location="West Europe"
    $VNetName="PacktPublishing"
    $ProdSubnetName="Production"
    $FWSubnetName="AzureFirewallSubnet"
    $FWpipName="PacktFirewall-PIP"
    $FWname="PackFirewall"
    ```

2. Create a resource group:

    ```
    New-AzResourceGroup -Name $RGName -Location $Location
    ```

3. Create subnets:

    ```
    $ProdSubnet=New-AzVirtualNetworkSubnetConfig '
    -Name $ProdSubnetName -AddressPrefix 172.16.0.0/24
    $FWSubnet=New-AzVirtualNetworkSubnetConfig '
    -Name ' $FWSubnetName -AddressPrefix 172.16.1.0/26
    ```

4. Create a virtual network containing two subnets:

```
$VNet=New-AzVirtualNetwork -Name $VNetName '
-ResourceGroupName $RGName -Location $Location '
-AddressPrefix 172.16.0.0/16 -Subnet
$ProdSubnet,$FWSubnet
```

5. Create a public IP address for a firewall:

```
$FWpip = New-AzPublicIpAddress -Name $FWpipName '
-ResourceGroupName $RGName -Location $Location '
-AllocationMethod Static -Sku Standard
```

6. Create a firewall:

```
$Azfw = New-AzFirewall -Name $FWname '
-ResourceGroupName $RGName -Location $Location '
-VirtualNetworkName $VNetName -PublicIpName $FWpipName
```

How it works...

Creating an Azure firewall using PowerShell is like creating an Azure firewall in the Azure portal. First, we defined variables containing names of the resources that will be used in later commands. Prior to creating a virtual network, we created two virtual network subnets, one of them named `AzureFirewallSubnet`, a requirement for deploying a firewall. Then, we created a virtual network containing two previously defined subnets and proceeded with creating a public IP address. In the last step, we created an Azure firewall using an IP address created as a public endpoint.

Creating an Azure firewall policy

An **Azure firewall policy** is an Azure resource that defines *application, network*, and *NAT rule* collections, as well as additional settings such as *TLS inspection*, **Intrusion Detection and Prevention System (IDPS)**, *threat intelligence*, and *DNS* settings. Azure Firewall Premium SKU uses an Azure firewall policy that allows the central management of firewalls via Azure Firewall Manager. In this recipe, you will create an Azure firewall policy, Premium tier, using the Azure portal.

Getting ready

Open a web browser and navigate to https://portal.azure.com.

How to do it...

To create an Azure firewall policy, complete the following steps:

1. In the Azure portal, open **Firewall Manager** and navigate to **Azure Firewall Policies**.

2. Select + **Create Azure Firewall Policy**.

3. Choose a subscription and resource group.

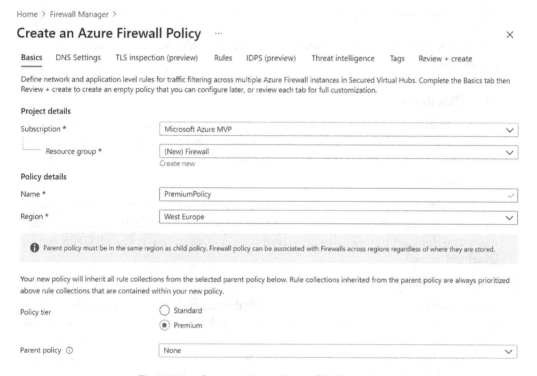

Figure 8.4 – Create an Azure Firewall Policy – Basics

4. Type the name of the policy and choose the Region.

5. For **Policy tier**, select **Premium**.

6. The new policy will inherit rule collection from a parent policy if you associate it with an existing policy. Leave the **Parent Policy** setting to **None**.

7. Then, click **DNS Settings**.

8. By default, **DNS Proxy** is **Disabled** and an Azure firewall uses Azure DNS. If **Enabled**, you can specify **Custom DNS Servers** and enable an Azure firewall to forward DNS requests to DNS servers specified in the **Custom DNS Servers** field. Leave the setting at **Disabled**.

9. Then, click **TLS Inspection**.

10. If **Enabled**, the TLS termination and inspection function is available. To use TLS inspection, you must provide a valid certificate stored in Azure Key Vault, and a managed identity with read permissions to secrets and certificates, as defined in a Key Vault access policy. If you do not have these resources ready, leave the **TLS Inspection** setting on **Disabled**.

> **Note**
>
> TLS 1.2 is supported as TLS 1.0, while TLS 1.1 will be deprecated and no longer supported.

11. Then, click **Rules**.

12. On the **Rules** page, you can add a rule collection or import rules from an Azure firewall. An Azure firewall supports NAT, network, and application rules. It is possible to configure rules after you create a policy.

13. Then, click **IDPS**.

14. On the **IDPS** page, you can configure **Policy** to the **Alert** or **Alert and deny** setting. The **Alert** setting will generate an alert when suspicious traffic is detected, while the **Alert and deny** setting will also deny traffic if a matching rule is found. **IDPS** can be **Disabled** as well.

15. Then, click **Threat intelligence**.

16. **Threat intelligence mode** can be in a **Disabled** state, and it also has two other options: **Alert Only** and **Alert and deny**. This capability will alert and block inbound and outbound network traffic to known malicious IP addresses and domains. The **Add allow list addresses** option allows you to specify that the **IP address, range, or subnet** or **Fqdn** fields can be excluded from **Threat intelligence** actions.

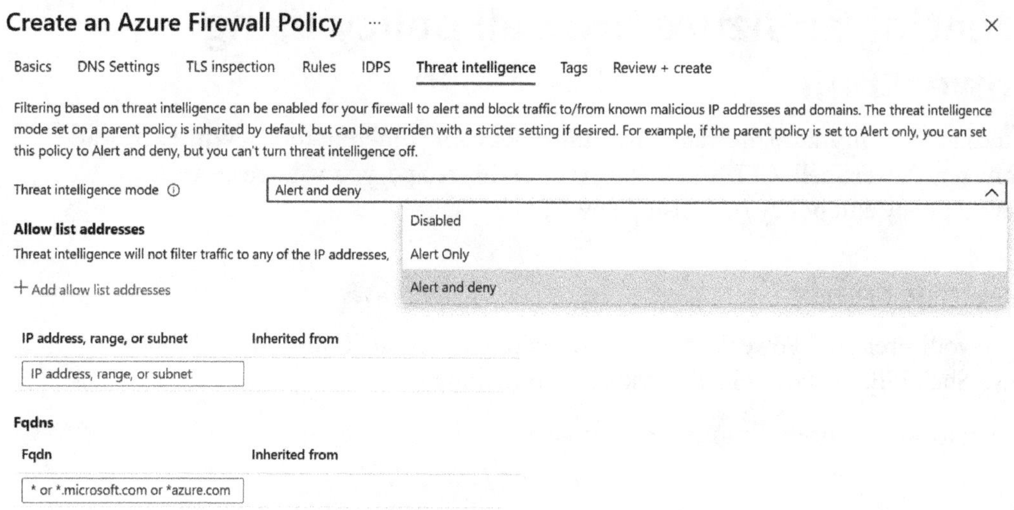

Create an Azure Firewall Policy   ...   ✕

Basics   DNS Settings   TLS inspection   Rules   IDPS   **Threat intelligence**   Tags   Review + create

Filtering based on threat intelligence can be enabled for your firewall to alert and block traffic to/from known malicious IP addresses and domains. The threat intelligence mode set on a parent policy is inherited by default, but can be overridden with a stricter setting if desired. For example, if the parent policy is set to Alert only, you can set this policy to Alert and deny, but you can't turn threat intelligence off.

Threat intelligence mode ⓘ   Alert and deny   ⌃

Allow list addresses   Disabled

Threat intelligence will not filter traffic to any of the IP addresses,   Alert Only

╋ Add allow list addresses   Alert and deny

IP address, range, or subnet   Inherited from

    IP address, range, or subnet

Fqdns

Fqdn   Inherited from

    * or *.microsoft.com or *azure.com

Figure 8.5 – Create an Azure Firewall Policy – Threat intelligence

17. Then, click **Tags**, where you can add tags to the Azure policy.

18. Click **Review + create** to validate the settings.

19. Click **Create** to create a firewall policy.

How it works...

An **Azure firewall policy** is an Azure resource that defines *application, network,* and *NAT* rules and rule collections, along with **threat intelligence** settings. An Azure firewall policy can be used as a standalone policy, or it can inherit settings from a *parent policy*.

You can create and associate firewall policies with multiple firewalls using **Firewall Manager**. If you have deployed **virtual WAN**, you can associate an Azure firewall and firewall policy with **Virtual WAN Hub** to make a **secured virtual hub**, or with a virtual network, making a **hub virtual network or secured virtual network**.

Creating an Azure firewall policy using PowerShell

You can create an Azure firewall policy and rules using different tools: **Azure Firewall Manager**, **PowerShell**, **CLI**, and **REST API**. In this recipe, you will create an Azure firewall policy and policy rules using PowerShell.

Getting ready

Open your preferred PowerShell tool – this could be Visual Studio Code, Windows PowerShell ISE, the PowerShell console, or something else.

Sign in to your Azure account: `Connect-AzAccount`.

How to do it...

To create an Azure firewall policy and policy rules using PowerShell, complete the following steps:

1. Define firewall policy resources:

   ```
   $RGName="Firewall"
   $Location="West Europe"
   $fwPolicyName="FW-policy"
   $netCollName="NetworkCollectionGroup"
   $netRuleName="AllowGoogleDNS"
   $appCollName="AppCollectionGroup"
   $appRuleName="Allow-Packt"
   ```

2. Create a resource group:

   ```
   New-AzResourceGroup -Name $RGName -Location $Location
   ```

3. Create a firewall policy:

   ```
   $FWpolicy = New-AzFirewallPolicy -Name $fwPolicyName '
   -ResourceGroupName $RGName -Location $Location
   ```

4. Create a network rule collection group:

   ```
   $nrRCGroup = New-AzFirewallPolicyRuleCollectionGroup '
   -Name $netCollName -Priority 1200 '
   -FirewallPolicyObject $FWpolicy
   ```

5. Configure a firewall policy network rule. The network rule allows outbound access to two Google DNS servers at port 53 (DNS):

```
$netRule01 = New-AzFirewallPolicyNetworkRule '
-name $netRuleName -protocol UDP '
-sourceaddress 172.17.0.0/24 '
-destinationaddress 8.8.8.8,8.8.4.4 -destinationport 53
```

The following screenshot shows the firewall policy network rule setting in the Azure portal as a result of completing the PowerShell command:

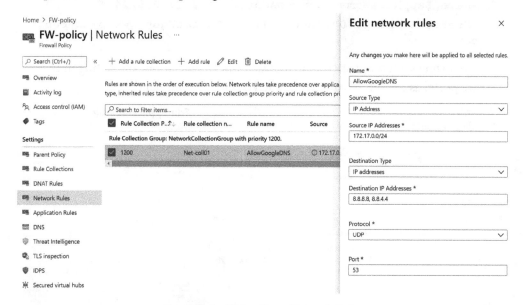

Figure 8.6 – FW-policy – Network Rules

6. Configure the network rule collection:

```
$netColl01 = New-AzFirewallPolicyFilterRuleCollection '
-Name Net-coll01 -Priority 1200 -Rule $netRule01 '
-ActionType "Allow"
```

7. Associate the network rule collection group and the network rule collection with a firewall policy:

```
Set-AzFirewallPolicyRuleCollectionGroup '
-Name $nrRCGroup.Name -Priority 1200 '
-RuleCollection $netColl01 -FirewallPolicyObject
$FWpolicy
```

8. Create an application rule collection group:

```
$arRCGroup = New-AzFirewallPolicyRuleCollectionGroup '
-Name $appCollName -Priority 1300 '
-FirewallPolicyObject $FWpolicy
```

9. Configure a firewall policy application rule. The application rule allows outbound access to www.packtpub.com:

```
$appRule01 = New-AzFirewallPolicyApplicationRule '
-Name $appRuleName -SourceAddress 172.17.0.0/24 '
-Protocol "http:80","https:443" -TargetFqdn www.packtpub.
com
```

The following screenshot shows the firewall policy application rule setting in the Azure portal as a result of completing the PowerShell command:

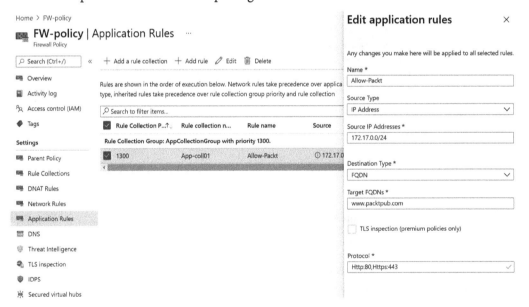

Figure 8.7 – FW-policy – Application Rules

10. Configure the application rule collection:

```
$appColl01 = New-AzFirewallPolicyFilterRuleCollection '
-Name App-coll01 -Priority 1300 -Rule $appRule01 '
-ActionType "Allow"
```

11. Associate the application rule collection group and the application rule collection with a firewall policy:

```
Set-AzFirewallPolicyRuleCollectionGroup '
-Name $arRCGroup.Name -Priority 1300 '
-RuleCollection $appColl01 -FirewallPolicyObject
$FWpolicy
```

12. Open **Firewall Policies,** select the firewall policy for which you have defined firewall rules, and select **Rule Collections**. Observe the rule collections you have created using PowerShell:

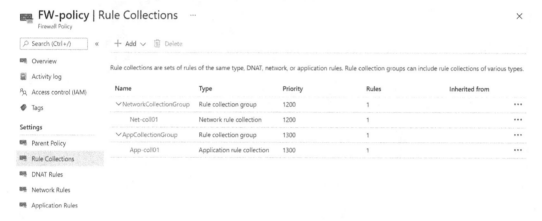

Figure 8.8 – FW-policy – Rule Collections

How it works...

To create an Azure firewall policy using PowerShell, first, we defined the resource variables and created a resource group. But, before anything else, we created a firewall policy and proceeded to create *rules, rule collections, and rule collection groups.*

Rule collection groups contain one or more *rule collections*, and *rule collections* contain one or more *rules.*

We created a *network rule collection group*, containing a *network rule collection*, which contains one *network rule*. Later, we created an *application rule collection group*, containing an *application rule collection*, which contains one *application rule*.

9
Information Protection

Users and organizations work with ever-growing volumes and types of data, spread over various repositories, and security personnel and administrators have to protect an increasing number of resources. Not all resources are equal; some store more sensitive information than others, and to effectively protect their environments, security professionals have to know where to prioritize the protection.

Microsoft Defender for Cloud supports SQL information protection policies – a classification mechanism, and Microsoft Purview integration as well, which is a data governance service. With these capabilities and service integrations, Defender for Cloud can provide additional alerts and recommendations based on discovered and monitored data sensitivity types.

In this chapter, you will learn how to work with sensitivity labels and information types as well as how to work with data classification. This will provide valuable information to Defender for Cloud, enabling it to generate alerts and recommendations based on information policy data.

We will cover the following recipes in this chapter:

- Creating and managing sensitivity labels
- Creating and managing information types and managing information protection policy

To complete the recipes in this chapter, the following is required:

- An Azure subscription
- A web browser, preferably Microsoft Edge
- Defender for Cloud plans
- Resources in an Azure subscription, such as virtual machines, storage, SQL server, and Logic Apps. Defender for Cloud will create resource recommendations based on available resources.

The code samples can be found at `https://github.com/PacktPublishing/Microsoft-Defender-for-Cloud-Cookbook`.

Please note: To display SQL Information Protection button in Defender for Cloud, you must have tenant-level permissions. Alternatively, you can access Information Protection blade completing following steps:

1. Open **Azure SQL** blade.
2. Select a database.
3. On the menu, under **Security** section, click **Data Discovery & Classification**
4. Click **Configure**.

Creating and managing sensitivity labels

Sensitivity labels are one of the two foundations of SQL information protection and data classification mechanisms.

Labels define a degree or a level of data sensitivity and constitute a core classification attribute. Although information protection policy options contain predefined labels, it is important to create labels that match your data sensitivity.

In this recipe, you will learn how to create and manage sensitivity labels in SQL Information Protection.

Getting ready

Open a web browser and navigate to `https://portal.azure.com`.

How to do it...

To create a new sensitivity label in SQL Information Protection, in Defender for Cloud, complete the following steps:

1. In the Azure portal, open **Microsoft Defender for Cloud**.

2. On the left-hand menu, click **Environment settings**.

3. On the top menu, click **SQL Information Protection**.

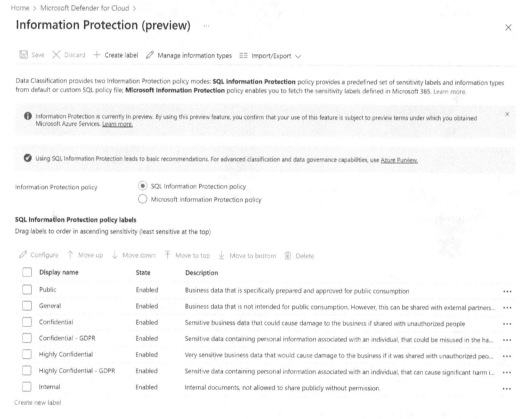

Figure 9.1 – Information Protection

SQL data classification supports two Information Protection policies: **SQL Information Protection policy** and **Microsoft Information Protection policy**. In this example, we will use the SQL Information Protection policy. Select **SQL Information Protection Policy**.

4.   From the top menu, click the **+ Create label** button.

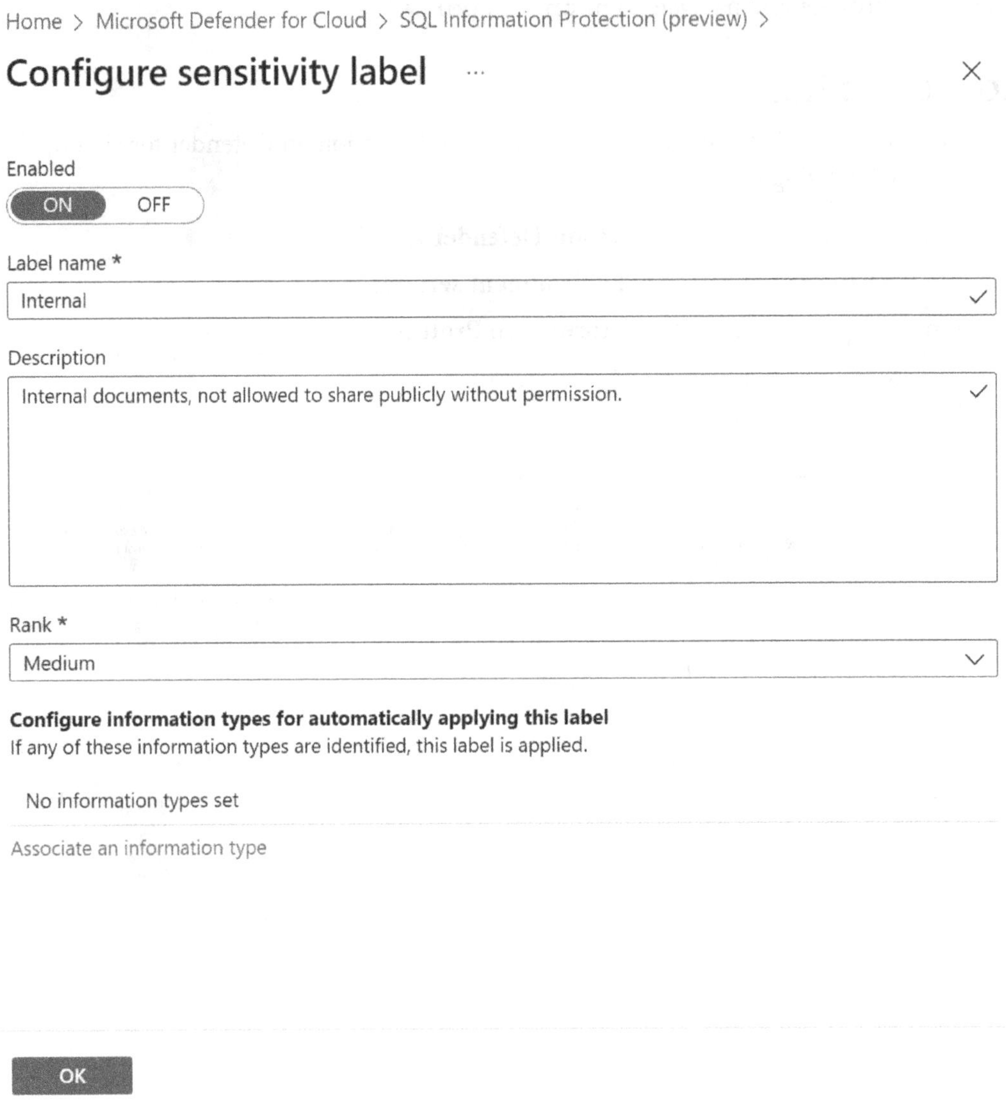

Home > Microsoft Defender for Cloud > SQL Information Protection (preview) >

Configure sensitivity label   ...   ×

Enabled

[ON   OFF]

Label name *

| Internal   ✓ |

Description

| Internal documents, not allowed to share publicly without permission.   ✓ |

Rank *

| Medium   ⌄ |

Configure information types for automatically applying this label
If any of these information types are identified, this label is applied.

No information types set

Associate an information type

OK

Figure 9.2 – Configure sensitivity label

5.   Sensitivity labels can be **Enabled** or **Disabled**. To start creating a sensitivity label and finish the process later, or to create a sensitivity label but not use it right away, select **Disabled**. Select **Enabled** to be able to use the label later.

6. Type a label name. Label names should reflect the sensitivity of the content that will be labeled, such as Confidential, Top secret, Internal, and similar. Avoid naming labels Financial, Research, Contracts, and the like, as these words do not necessarily describe the sensitivity of the content.

7. Type a label description that will accurately describe what kind of content is being labeled.

8. From a drop-down menu, choose **Rank**. Rank describes a degree of content sensitivity.

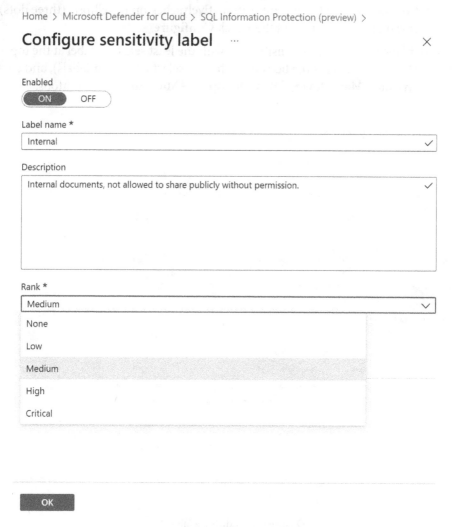

Figure 9.3 – Configure sensitivity label rank

9. The **Configure Information types for automatically applying this label** section allows you to define an information type to automatically apply this label. To choose a predefined information type, you would click on the **Associate an information type** link. You will not do that at this time as we will create a new information type and associate it with this label later.

Click **OK** to finish creating a label.

10. A newly created label is shown in the list of available labels.

11. To edit and change the label settings, on the menu under **SQL Information Protection labels**, click **Configure**. Alternatively, click on the ellipsis (three dots) on a label row on the right side and then click **Configure**.

12. To order labels in ascending sensitivity – with the least sensitive label at the top and the most sensitive label at the bottom of the list, select a label (or labels), and click the ↑**Move up**, ↓**Move down**, ⊤**Move to top**, or ⊥**Move to bottom** buttons.

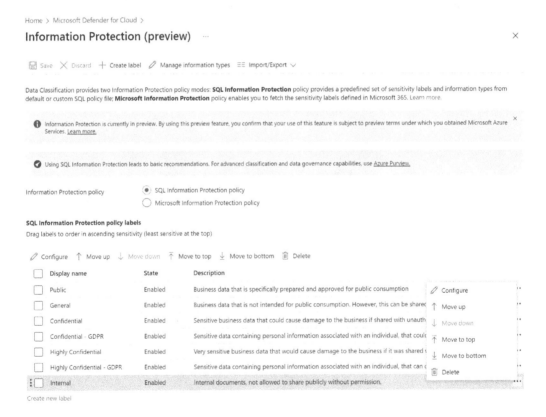

Figure 9.4 – Editing a label

How it works...

The labels you specify in your SQL Information Policy, or Information Protection policy, define data sensitivity levels and are used in data discovery, classification, labeling, and reporting mechanisms.

There's more...

Alternatively, you can also access the SQL Information Protection blade on the SQL database Azure portal pages. To access SQL Information Protection via the SQL database blade, complete the following steps:

1. In the Azure portal, open **SQL Database**.
2. On the left-hand menu, in the **Security** section, select **Data Discovery & Classification**.
3. On the top menu, click **Configure**.

Creating and managing information types and managing information protection policy

Information types are the second of the two foundations of SQL Information Protection and data classification mechanisms. Information types are used to provide supplementary and more granular details of the discovered and classified data.

In this recipe, you will learn how to create and manage an information type that can be used in the SQL Information Protection policy. Moreover, you will also learn how to import and export an information protection policy.

Getting ready

Open a web browser and navigate to `https://portal.azure.com`.

How to do it...

To create and manage information types, complete the following steps:

1. In the Azure portal, open **Microsoft Defender for Cloud**.
2. On the top menu, click **SQL Information Protection**.

3. In the **Information Protection** blade, on the top menu, click **Manage Information Types**.

Home > SQL Information Protection (preview) >

Information types   ...

+ Create information type

Create and manage information types
Drag information types to order in ascending discovering ranking

🖉 Configure   ↑ Move up   ↓ Move down   ↑ Move to top   ↓ Move to bottom   🗑 Delete

	Information type	State	Associated label	Type
☐	Networking	Enabled	Confidential	Built-in
☐	Contact Info	Enabled	Confidential	Built-in
☐	Credentials	Enabled	Confidential	Built-in
☐	Name	Enabled	Confidential - GDPR	Built-in
☐	National ID	Enabled	Confidential - GDPR	Built-in
☐	SSN	Enabled	Confidential - GDPR	Built-in
☐	Credit Card	Enabled	Confidential	Built-in
☐	Banking	Enabled	Confidential	Built-in
☐	Financial	Enabled	Confidential	Built-in
☐	Health	Enabled	Confidential - GDPR	Built-in
☐	Date Of Birth	Enabled	Confidential - GDPR	Built-in
☐	Other	Enabled	Confidential	Built-in

Create new information type

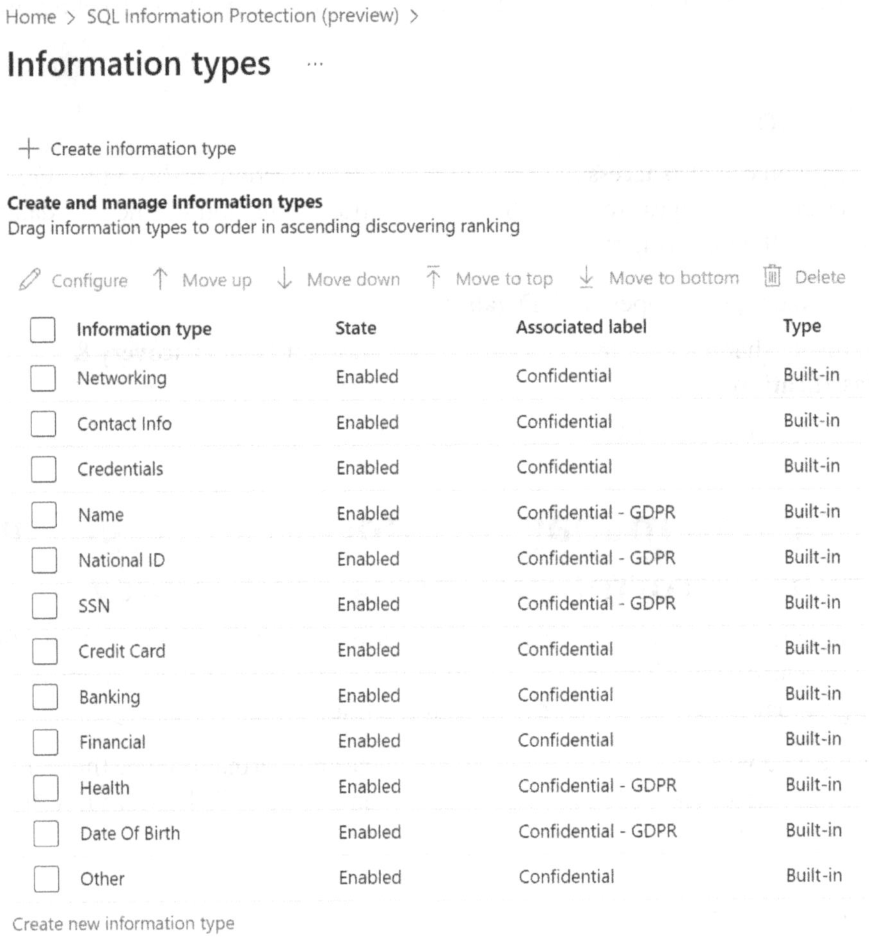

Figure 9.5 – Information types

4. The **Information types** blade contains a list of predefined and custom information types and their associated data:

I. **Name** – Information type

II. **State** – Enabled or disabled

III. **Associated label**

IV. **Type** – Built-in or Custom

On the top menu, click + **Create information type**.

5. On the right, the **Configure information type** window opens. Similarly to when configuring a sensitivity label, here as well you have the option to enable or disable the information type you are creating. Select **Enabled**:

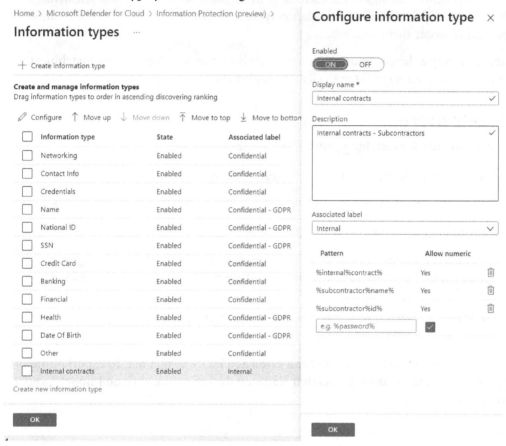

Figure 9.6 – Configuring information types

6. In the **Display name** field, type a name for the information type you are configuring. The name should be concise, simple, and easy to understand.

7. In the **Description** field, type a description of an information type you are configuring. Describe accurately what information type you are defining.

8. In the **Associated label** menu, choose an information label to associate with the current information type. Alternatively, editing an information label gives you the option to associate it with an information type.

9. Now, you must define an information type pattern. In the **Pattern** field, enter a pattern to match the data.

10. Click **OK**.

How it works...

Information types provide additional details relating to sensitivity labels since search patterns use defined strings to match content in databases and classify it as sensitive. While you can't delete built-in information types, you can create your own information types and associate them with labels.

In the same way as labels, information types use list order to prioritize data matching: the higher an information type is in the list, the higher the matching priority.

There's more...

Use standard, regular matching patterns such as the following:

- % (percent) – to match any string of zero or more characters
- _ (underscore) – to match any single character in a string

For more information, refer to the following links:

- `https://docs.microsoft.com/en-us/sql/t-sql/language-elements/string-operators-transact-sql?view=sql-server-ver15`
- `https://dev.mysql.com/doc/mysql-tutorial-excerpt/5.7/en/pattern-matching.html`

Information policies can be exported and saved as a JSON format file. To export or import a policy, on the **Information Protection** blade, on the top menu, click on **Import/Export** and then select the desired option:

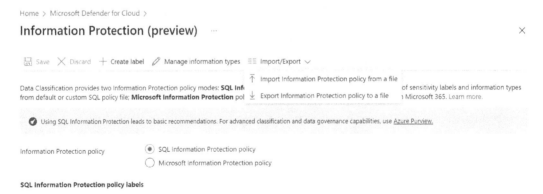

Figure 9.7 – Importing and exporting an Information Protection policy

As with any other JSON file, the policies can be edited and the file can be imported into a new Information Protection policy, overwriting the current policy:

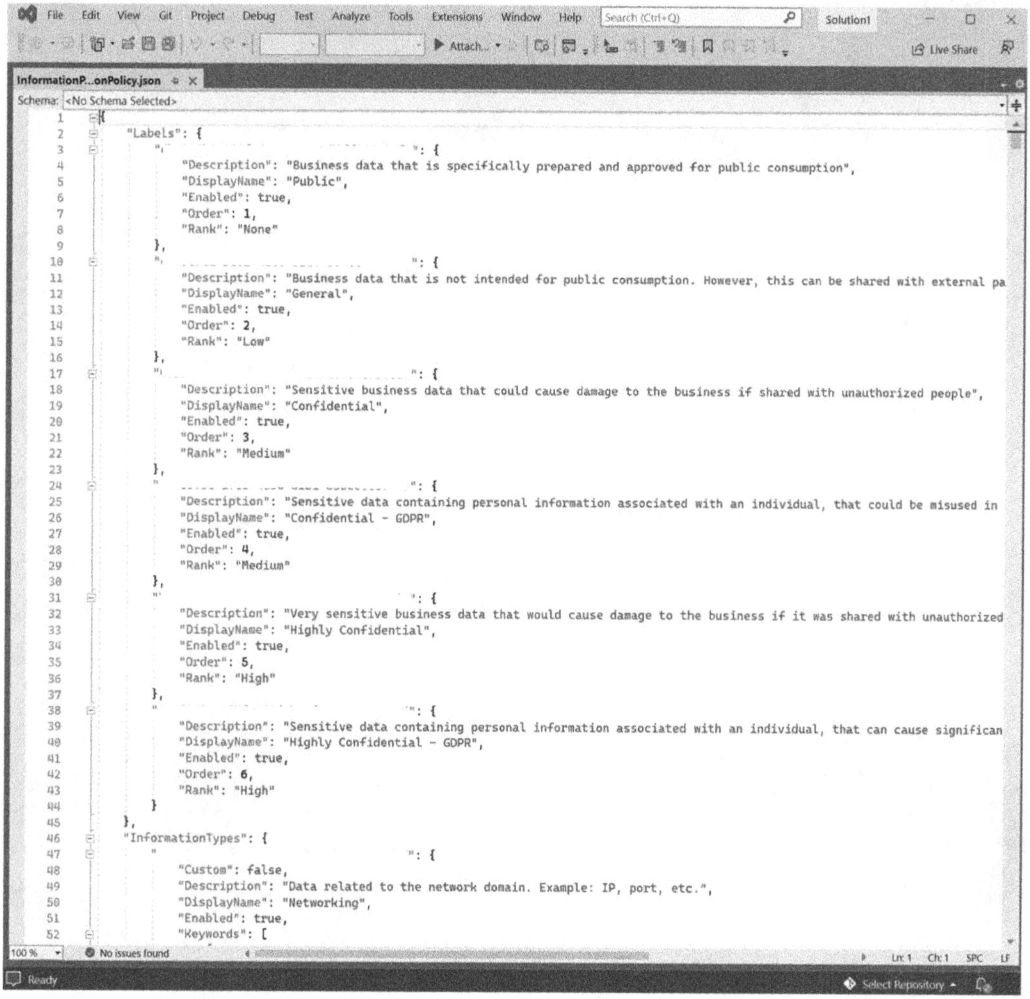

Figure 9.8 – InformationProtectionPolicy.json file

10

Workbooks

In this chapter, you will learn how to create and manage workbooks in Microsoft Defender for Cloud. You can create workbooks in several ways and, once created, you can change and modify them as you need, as well as share and redeploy them as needed in other tenants.

Workbooks provide a way to analyze data, gain insight into Azure data and trends, discover baselines, create powerful visual reports, and much more.

We will not explain how to create a workbook in detail—such as adding parameters, links, queries, and other workbook elements—as this is out of the scope of this book. We will show you how to get started with creating workbooks, and then you can take it from there and modify workbooks further to suit your needs.

We will cover the following recipes in this chapter:

- Creating a workbook from an existing template
- Creating a workbook from an empty workbook
- Managing workbooks and workbook templates

Technical requirements

To successfully complete the recipes in this chapter, the following is required:

- An Azure subscription

- A web browser, preferably Microsoft Edge

- Microsoft Defender for Cloud plans

- Resources in an Azure subscription, such as **virtual machines** (**VMs**), storage, a **Structured Query Language** (**SQL**) server, and Logic Apps. Microsoft Defender for Cloud will create resource recommendations based on available resources.

The code samples can be found at `https://github.com/PacktPublishing/Microsoft-Defender-for-Cloud-Cookbook`.

Creating a workbook from an existing template

You can create Microsoft Defender for Cloud workbooks in several ways, and using a pre-created or existing workbook template is one way to create an additional workbook.

In this recipe, you will learn how to create a workbook from an existing template.

Getting ready

Open a web browser and navigate to `https://portal.azure.com`.

How to do it...

To create a workbook from an existing template, complete the following steps:

1. In the Azure portal, open **Microsoft Defender for Cloud**.
2. On the left menu, click **Workbooks**.

3.   Click on a template you want to edit. You will create a new workbook based on the selected template. In the following example, we will select the **Compliance Over Time** workbook:

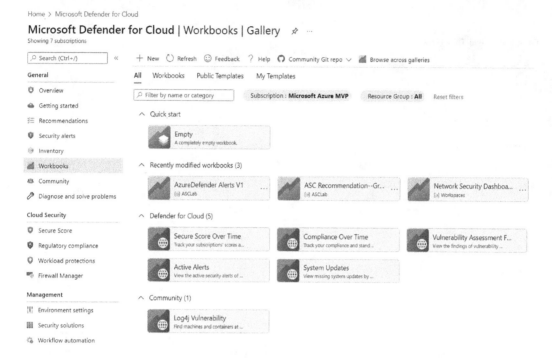

Figure 10.1 – Workbooks gallery

4. From the drop-down menus, select values for **Workspace** and **Subscription**, and for **Standard name**, select one or more regulatory compliance standards. In this example, select the **Azure Security Benchmark** standard:

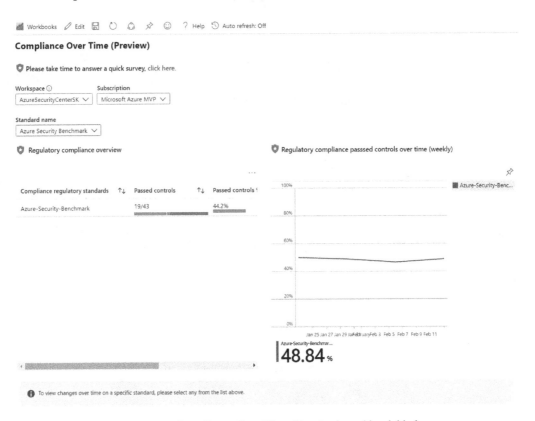

Figure 10.2 – Compliance Over Time (Preview) workbook blade

5. The top menu on the **Workbooks** page contains a command for managing workbooks. To make changes to a workbook and to save it under a different name, click **Edit**.

6. When you selected **Edit**, you changed from reading mode to editing mode, and the menu button changes accordingly to **Done Editing**. You can distinguish between the two modes by looking at the right side of a blade and identifying the ↑ **Edit** button.

7. Click on the ↑ **Edit** button next to a workbook title:

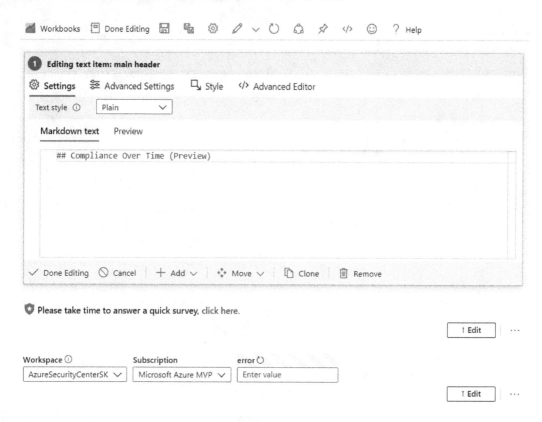

Figure 10.3 – Editing text item: main header

8. Change the workbook title to `Azure Security Benchmark Compliance Over Time`.

9. Click **Done Editing**.

10. On the lower-right part of the **Standard name** page, click on the ↑ **Edit** button:

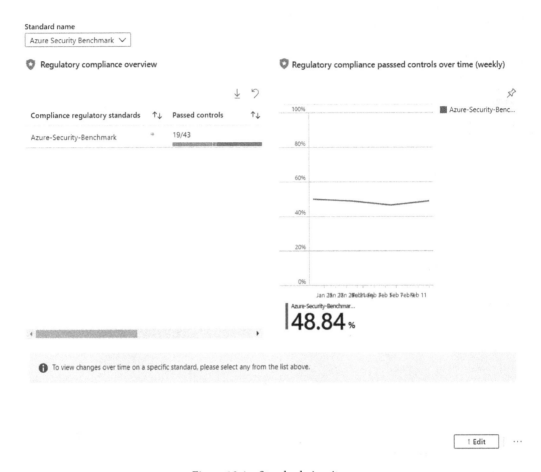

Figure 10.4 – Standard view item

11. Inside ❹ **Editing group item: Standard view**, next to the **Standard name** dropdown, click on the ↑ **Edit** button:

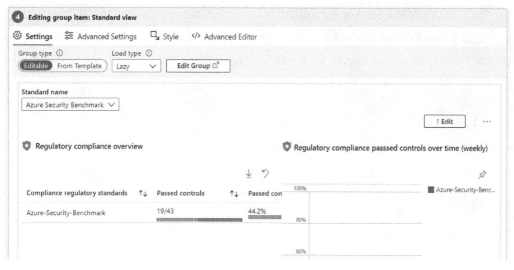

Figure 10.5 – Workbook editor: Standard view

12. Inside the **❶ Editing parameters item: parameters – 0** field, change the text in the **Display name** field from **Standard name** to `Regulatory Compliance Standard Name`:

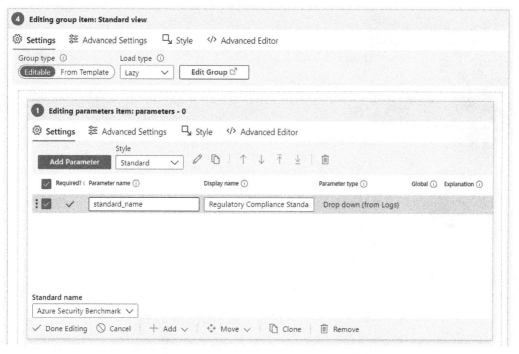

Figure 10.6 – Workbook editor: editing parameters

13. Click **Done Editing** in the ❶ **Editing parameters item: parameters – 0** field.

14. Click **Done Editing** in the ❹ **Editing group item: Standard view** field.

15. On the top workbook menu, click on the **Save as** icon (two floppy disks, next to a cog icon):

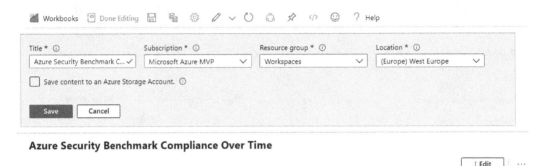

Figure 10.7 – Workbook save dialog box

16. In the **Title** field, type a new workbook name. Choose a subscription, resource group, and location to save the workbook. Click **Save**.

17. Return to the **Workbooks** blade. Select the **All** tab to display all workbooks:

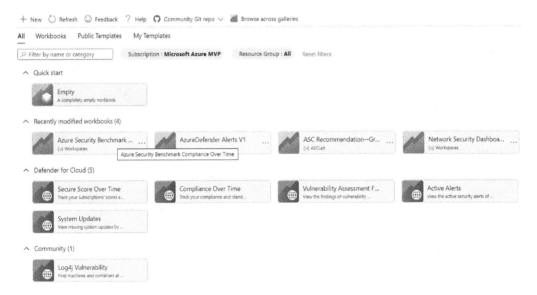

Figure 10.8 – Microsoft Defender for Cloud: Workbooks

18. Under **Recently modified workbooks**, identify the recently modified **Azure Security Benchmark Compliance Over Time** workbook and select **Azure Security Benchmark Compliance Over Time workbook**.

19. Observe a workbook you have created from an existing workbook template.

How it works...

Microsoft Defender for Cloud at this time includes out-of-the-box workbooks, such as **Secure Score Over Time, Compliance Over Time, Vulnerability Assessment Findings, Active Alerts**, and **System Updates**. On certain occasions, editing and modifying an existing template can save time. Many workbook templates can be found at `https://github.com/Azure/Microsoft-Defender-for-Cloud/tree/main/Workbooks`.

Creating a workbook from an empty workbook

Microsoft Azure supports creating workbooks from scratch or from a blank workbook. This gives you the possibility to take advantage of the powerful workbook editor and its capability to define the look and feel of a new workbook in detail.

In this recipe, you will learn how to create a workbook from an empty workbook.

Getting ready

Open a web browser and navigate to `https://portal.azure.com`.

How to do it...

To create a workbook from an empty workbook, complete the following steps:

1. In the Azure portal, open **Microsoft Defender for Cloud**.

2.   On the left menu, click **Workbooks**:

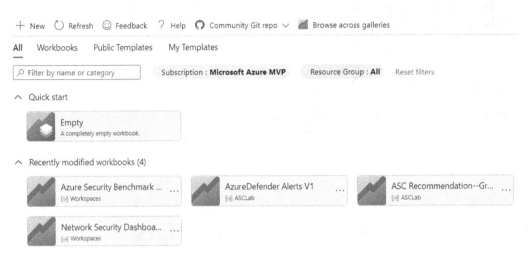

Figure 10.9 – Workbooks blade

3.   On the top menu is the **+ New** button, and the **Quick start** section contains an **Empty** button. Click on either button to start creating a workbook from scratch— that is, an empty workbook.

4.   An empty workbook editor opens. If you were to develop and create a workbook from scratch, you would have to add text, parameters, links, queries, and metrics, possibly in groups:

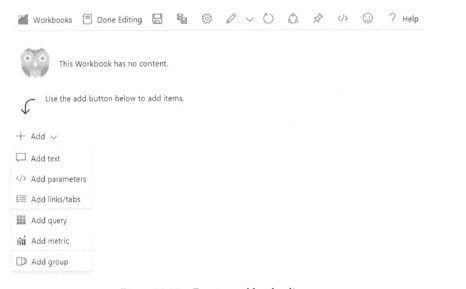

Figure 10.10 – Empty workbook editor

In this exercise, you will use a saved template and create your version of a workbook. On the toolbar, click the </> button to access **Advanced Editor**.

5. Open a new browser tab and navigate to `https://github.com/Azure/Microsoft-Defender-for-Cloud/blob/main/Workbooks/Regulatory%20Compliance/regulatorycompliance.workbook`.

6. Copy the workbook content to the clipboard.

7. Select **Gallery Template** for the **Template Type** field. You would use **Gallery Template** to create a template for the workbook gallery, and **ARM Template** to create a template usable with **Azure Resource Manager** (**ARM**).

8. Paste and replace the gallery template content with the workbook content copied in *Step 6*:

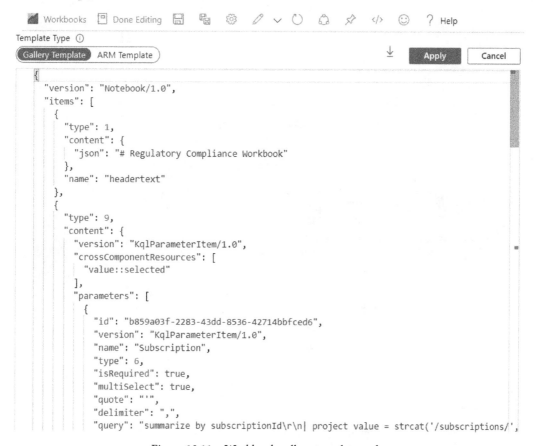

Figure 10.11 – Workbook gallery template code

9. Click **Apply**. As you do so, **Advanced Editor** closes, and you will see the result of a workbook template in the editor:

Figure 10.12 – Editing workbook fields

10. In the editor, click on the ↑ **Edit** button to edit and modify desired workbook fields. Click **Done Editing** when you are done with editing the workbook to exit the workbook editor.

11. Click the **Save** button:

Regulatory Compliance Workbook

Subscription: All ∨

This workbook visualises the Regulatory Compliance standards and findings from Azure Defender.

Overview   Regulatory Compliance Detail

Figure 10.13 – Save workbook dialog

12. Enter the workbook title, subscription, resource group, and location to save the workbook, and click **Save**.

How it works...

Regardless of the many workbook templates available at `https://github.com/Azure/Microsoft-Defender-for-Cloud/tree/main/Workbooks`, you might want to start with a blank workbook to be able to design it fully to your specifications, designs, and desires, without unnecessary editing and changing existing code.

There's more...

To write your own workbooks, you should be familiar with **Kusto Query Language** or **KQL**, the query language required not only to build custom workbooks but also for data analysis, viewing monitor data from multiple resources, creating textual and visual reports, hunting for threats, and much more.

Here are some references to online resources that might help you to learn about workbooks and KQL:

- *Azure Monitor Workbooks*

  `https://docs.microsoft.com/en-us/azure/azure-monitor/`
  `visualize/workbooks-overview`

- *Kusto Query Language (KQL) overview*

  `https://docs.microsoft.com/en-us/azure/data-explorer/`
  `kusto/query/`

- *Write your first query with Kusto Query Language*

  `https://docs.microsoft.com/en-us/learn/modules/write-`
  `first-query-kusto-query-language/`

- *SC-200: Create queries for Microsoft Sentinel using Kusto Query Language (KQL)*

  `https://docs.microsoft.com/en-us/learn/paths/sc-200-`
  `utilize-kql-for-azure-sentinel/`

- *Must Learn KQL*

  `https://aka.ms/MustLearnKQL`

- *Kusto Query Language (KQL) from Scratch*

  `https://www.pluralsight.com/courses/kusto-query-language-`
  `kql-from-scratch`

Managing workbooks and workbook templates

Once you create a workbook, you can modify its parameters, rename, delete, and lock a workbook, and perform well-known Azure tasks. Additionally, you can deploy a workbook from a GitHub repository too.

In this recipe, you will learn how to manage workbooks and workbook templates and to deploy a workbook from a GitHub repository.

Getting ready

Open a web browser and navigate to `https://portal.azure.com`.

How to do it...

To manage workbooks and workbook templates in Microsoft Defender for Cloud and to deploy a workbook from a GitHub repository, complete the following steps:

1. In the Azure portal, open **Microsoft Defender for Cloud**.

2. On the left menu, click **Workbooks**:

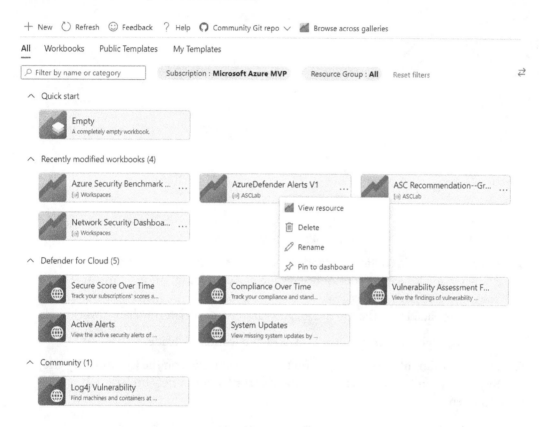

Figure 10.14 – Microsoft Defender for Cloud workbooks gallery

3. On the **Workbooks gallery** blade, under the **Recently modified workbooks** section, either click on an ellipsis on a workbook tile or right-click on a workbook tile to open a menu. The menu allows you to perform several actions on a workbook: **View resource**, **Delete**, **Rename**, or **Pin to dashboard**.

4.   Click **View resource**. A workbook blade opens:

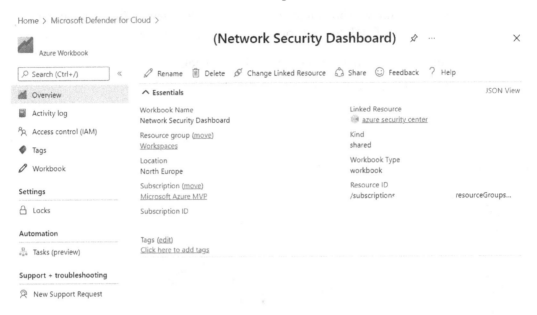

Figure 10.15 – Workbook blade

5.   A workbook blade contains a standard set of menus and groups of menus characteristic for every other Azure resource: **Activity log**, **Access control (IAM)**, **Tags**, and **Locks**, as well as a **Workbook** workbook-specific menu that contains workbook data. From the top menu, you can rename, delete, or share a workbook.

In the top-right corner, click on **X** to close the workbook blade.

6.   On the **Workbook Gallery** blade, on the top menu, click on the **Community Git Repo** menu and select **Microsoft Defender for Cloud**. A new browser tab will open.

Note: A new browser tab should open the `https://github.com/Azure/Microsoft-Defender-for-Cloud/tree/main/Workbooks` page. If not, open the page manually. This **Uniform Resource Locator** (**URL**) points to the Microsoft Defender for Cloud workbook gallery on GitHub.

7.   On the GitHub workbook gallery page, click on a workbook link that you want to deploy in Azure, in the Microsoft Defender for Cloud workbook gallery. In this example, we will install the **Defender for Cloud Coverage** workbook.

Click on the **Defender for Cloud Coverage** link:

Microsoft Defender for Cloud - Coverage Dashboard

Author: Tom Janetscheck

Microsoft Defender for Cloud plans are enabled per subscription what can make it hard to understand which plan is enabled on which subscription when you are not using policy enforcements. This workbook provides a consolidated view of Defender for Cloud coverage across all selected subscriptions.

Try it on the Azure Portal

You can deploy the workbook by clicking on the buttons below:

Figure 10.16 – Workbook GitHub details

8. Click on **Deploy to Azure** to deploy the workbook. Every workbook requires specific deployment parameters. Follow the onscreen instructions for the workbook instance to deploy the workbook. For example, specify **Subscription**, **Resource group**, and **Region** values, a workbook name, and other details. Click **Review + Create**, and then click **Create** to finish deploying the workbook:

Project details

Select the subscription to manage deployed resources and costs. Use resource groups like folders to organize and manage all your resources.

Subscription * ⓘ	Microsoft Azure MVP ⌄
Resource group * ⓘ	Workspaces ⌄
	Create new

Instance details

Region * ⓘ	(Europe) North Europe ✓
Workbook Display Name ⓘ	Defender for Cloud Coverage ✓
Workbook Type ⓘ	workbook ✓
Workbook Source Id ⓘ	Azure Security Center ✓
Workbook Id ⓘ	[newGuid()]

Figure 10.17 – Workbook deployment parameters

9. Return to the Microsoft Defender for Cloud workbooks gallery, and from the top menu, click **Refresh** to reload new content.

10. Click on the **Defender for Cloud Coverage** workbook.

11. Select one or more subscriptions to display the results.

How it works...

Business and technical requirements change over time and, accordingly, you need to edit, modify, change, and delete existing workbooks, as well as create new workbooks from newly created templates. Workbook editing in Microsoft Defender for Cloud provides convenient commands for all these actions.

Index

Subscribe to our online digital library for full access to over 7,000 books and videos, as well as industry leading tools to help you plan your personal development and advance your career. For more information, please visit our website.

Why subscribe?

- Spend less time learning and more time coding with practical eBooks and Videos from over 4,000 industry professionals

- Improve your learning with Skill Plans built especially for you

- Get a free eBook or video every month

- Fully searchable for easy access to vital information

- Copy and paste, print, and bookmark content

Did you know that Packt offers eBook versions of every book published, with PDF and ePub files available? You can upgrade to the eBook version at packt.com and as a print book customer, you are entitled to a discount on the eBook copy. Get in touch with us at customercare@packtpub.com for more details.

At www.packt.com, you can also read a collection of free technical articles, sign up for a range of free newsletters, and receive exclusive discounts and offers on Packt books and eBooks.

Other Books You May Enjoy

If you enjoyed this book, you may be interested in these other books by Packt:

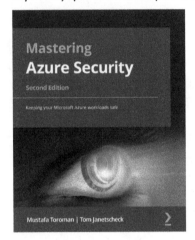

Mastering Azure Security - Second Edition

Mustafa Toroman, Tom Janetscheck

ISBN: 9781803238555

- Become well-versed with cloud security concepts
- Get the hang of managing cloud identities
- Understand the zero-trust approach
- Adopt the Azure security cloud infrastructure
- Protect and encrypt your data
- Grasp Azure network security concepts
- Discover how to keep cloud resources secure
- Implement cloud governance with security policies and rules

Penetration Testing Azure for Ethical Hackers

David Okeyode, Karl Fosaaen

ISBN: 9781839212932

- Identify how administrators misconfigure Azure services, leaving them open to exploitation
- Understand how to detect cloud infrastructure, service, and application misconfigurations
- Explore processes and techniques for exploiting common Azure security issues
- Use on-premises networks to pivot and escalate access within Azure
- Diagnose gaps and weaknesses in Azure security implementations
- Understand how attackers can escalate privileges in Azure AD

Packt is searching for authors like you

If you're interested in becoming an author for Packt, please visit `authors.packtpub.com` and apply today. We have worked with thousands of developers and tech professionals, just like you, to help them share their insight with the global tech community. You can make a general application, apply for a specific hot topic that we are recruiting an author for, or submit your own idea.

Share Your Thoughts

Now you've finished *Microsoft Defender for Cloud Cookbook*, we'd love to hear your thoughts! Scan the QR code below to go straight to the Amazon review page for this book and share your feedback or leave a review on the site that you purchased it from.

`https://packt.link/r/1-801-07613-8`

Your review is important to us and the tech community and will help us make sure we're delivering excellent quality content.